Flower Crops Cultivation and Management

Flower Crops Cultivation and Management

Ravi Kumar

RANDOM PUBLICATIONS
NEW DELHI (INDIA)

Flower Crops Cultivation and Management

ISBN 978-93-5111-951-7

Published in 2016 in India by

RANDOM PUBLICATIONS

4376-A/4B, Gali Murari Lal, Ansari Road
New Delhi-110 002
Phone : +9111-43580356, 011-23289044, 011-43142548
e-mail: sales@randompublications.com,
info@randompublications.com, randomexports@gmail.com

Reprinted 2018

Type Setting by : Friends Media, Delhi-110089

Digitally Printed at : Replika Press Pvt. Ltd.

Preface

Flower growing, once used to be a gardener's activity has today transformed into an astounding business. In India also, being an integral part of our rich heritage and culture, flower crops have become source of income as highly remunerative crops. Concerted efforts are being made in the country to boost the productivity, quality and export worthiness of several floricultural crops.

A flower, sometimes known as a bloom or blossom, is the reproductive structure found in flowering plants (plants of the division Magnoliophyta, also called angiosperms). The biological function of a flower is to effect reproduction, usually by providing a mechanism for the union of sperm with eggs. Flowers may facilitate outcrossing (fusion of sperm and eggs from different individuals in a population) or allow selfing (fusion of sperm and egg from the same flower). Some flowers produce diaspores without fertilization (parthenocarpy). Flowers contain sporangia and are the site where gametophytes develop. Flowers give rise to fruit and seeds. Many flowers have evolved to be attractive to animals, so as to cause them to be vectors for the transfer of pollen.

Some types of crops have flowers that may only be pollinated during a short period. If such a crop is not pollinated during that time, the flowers will fall and no seeds, berries or fruit will develop. There have to be sufficient numbers of bees in the pollinated crop. This is especially important in crops where the single flower may only be pollinated in a restricted time or in crops where the nectar production, or bee visits only take place during days where the temperature is at a certain level. In such a crop, the pollination in some years has to take place within three or four days. This can be the case in growing white clover seed.

The present book is designed for students taking courses in preparation for commercial, educational, or sales work in floriculture. It will also serve as a source book for operators and workers in the floriculture industry who may or may not have formal training in floriculture.

– *Author*

Contents

1

New Flower Crops

Most edible crops have been introduced into cultivation thousands of years ago. There are only a few new edible plants in the contemporary western horticulture, such as pecan, blueberry, and kiwifruit, but even these plants have been cultivated since ancient days by local farmers in their native region. This is not the case with ornamental crops.

Many of the commercial cut flowers and pot-plants grown today have not been cultivated commercially until several years ago. The ornamental plant industry is characterized by its great diversity. There are more ornamental species cultivated today than all other agricultural and horticultural crops combined.

Table. Quantities of Various Exportable Cut Flowers from Israel in the 1996/7 Export Season.

Flowers	Exportable Flowers (Millions of Stems)
Roses	453
Carnation	144
Gypsophila	116
Solidago	105
Ruscus	78
Wax flower	74
Hypericum	48
Gerbera	45
Limonium	41
Aster	35
Helianthus	32
Asclepias	27
Anemone	27
Safari sunset	20
Anigozanthos	17
Phlox	9
Others	210

In some ways the introduction of new ornamental crops is easier that of edible crops. Neither their nutritional value nor their general toxicity to human has to be considered, as evident in plants such as Aconitum, Diffenbachia,Oleander, and many others. Our main considerations in the introduction of new ornamental crops are the esthetic value, production costs, postproduction longevity, quality, and marketability.

The introduction of new crops includes many research stages, that start with the initial search and screening and concludes when the product is introduced commercially, as detailed in my other presentation in this proceedings (Halevy 1999). Ten years ago the traditional major crops constituted over 60 per cent of the cut flowers grown in and exported from Israel. This year over 60 per cent of the exportable flowers are "new crops," most of them have not been grown commercially 10 years ago. Many of these new commercial flower crops are not even mentioned in a recently published textbook on floriculture . There are several sources that serve for the introduction of new plant material as potential plant crops.

MINOR OUTDOOR GROWN CROPS

Many of the new greenhouse floral crops, grown and exported during the winter, are the so called "Summer Flowers." They are field grown plants that were used in Europe during their natural flowering season in the summer. Their introduction as a year round crop requires developing physiological and horticultural techniques for out of season production. Gypsophila and peony described above (Halevy 1999) are typical examples of such crops. Other examples are listed below.

ACONITUM NAPELLUS L., RANUNCULACEAE (MONK'S HOOD)

This is a tuberous plant native to Europe. For winter flowering, tubers are cold stored during the summer and pretreated with gibberellic acid before planting.

ASCLEPIAS TUBEROSA L., ASCLEPIADACEAE (BUTTERFLY WEED) AND A. INCARNATA L. (SWAMP MILKWEED)

Both plants are native to the US and considered as weeds there. They are absolute long day (LD) plants that require warm temperature during their growth and flowering. For winter production they are grown in heated greenhouses and provided with supplementary light at night.

ACHILLEA FILIPENDULINA LAM., ASTERACEAE (YARROW)

Native to East Asia it is used mainly for summer harvest as dry flowers. Year round production is obtained by digging the crowns and cold storing them for a few weeks before replanting.

LIATRIS SPICATA WILLD., ASTERACEAE

Native to Eastern US, for winter production, tubers are cold-stored during the summer and plants are lighted in the field.

PHLOX PANICULATA L., POLEMONIACEAE

Native to Eastern US, this herbaceous summer perennial is now grown for year round production in greenhouses. It is a LD plant, requiring supplementary night lighting.

SOLIDAGO SP. L., ASTERACEAE (GOLDENROD)

Species of goldenrod native to North America are considered as weeds there. New interspecific hybrids turned this plant into an important cut flowers. For winter production plants first receive LD to extend their stems and then are exposed to the natural winter short days (SD) for flower initiation and development.

TRACHELIUM CAERULEUM L., CAMPANULACEAE

Native to South Europe, it is an absolute LD plant and grown in the warmer parts of Israel for winter production.

MINOR FRUITS AND NUTS

Many types of fruits and nuts can be grown in Georgia due to our mild climate. Your county extension office can give you publications on fruits and nuts commonly grown in the state. This publication provides an outline of the culture and management of the exotic and uncommon fruits and nuts that can be grown in Georgia.

BANANA

Bananas (Musa spp.) are attractive landscape plants popular with many gardeners in Georgia. They may be grown with limited success outdoors in south Georgia. They usually require 12 to 18 months to produce a flower stalk. The fruit takes four to eight months to mature, depending on the temperature during the growing season. When winters are mild in south Georgia or when the plants are protected in north Georgia, the stalk may survive the winter and produce fruit in the second year. When winters are cold (below 25 degrees F), the tops are usually killed, so no fruit is produced. Some people grow bananas by potting the plants in tubs each fall and carrying them through the winter in a basement. Others build an insulating cage around the trunk using chicken wire and pine straw, enabling the plant to survive outdoors during the winter.

A third method of overwintering the plant is to remove most of the soil under the large rhizome but leave some roots attached on one side. Then lay the plant on the ground and cover both the rhizome and stalk with about 1 foot of soil. In the spring, after danger of freezing weather is past, remove the soil and stand the plant up and tie it to a fence post set near the base.

Bananas require plenty of water and fertilizer for good fruit production. Adjust soil pH to 5.5 to 6.5 before planting. Mulching is very beneficial because it conserves moisture, retards weeds, and protects the rhizomes from winter freezing. Water regularly and deeply during the summer if rainfall is poor. If the soil is extremely wet, root rot may develop. From late spring through summer, fertilize monthly with about 1 pound of 10-10-10 per mature plant cluster and 2 to 4 ounces of 10-10-10 per young plant. Scatter the fertilizer evenly in a large circle under the plants.

You need to prune bananas. In the beginning, let only one main stalk develop from each rhizome (underground bud). After six months, allow a replacement sucker to grow, because the main trunk is removed after fruiting. Unneeded suckers can be used to establish new plantings. The rhizomes may be dug up and divided for propagation of additional plants. The 'Cavendish' variety is probably the banana variety best adapted to Georgia.

The plant grows about 7 feet tall, produces good quality fruit, and is slightly more cold hardy than others. The 'Raja puri' variety is reputed to produce bananas in the first growing season. Plants set in the spring, 1998, in Savannah, Georgia, produced fruit, but it did not mature before frost. Additional information will be available in subsequent years. Many of the ornamental bananas grown in Georgia produce small fruit full of very hard, large seeds and usually are not edible. Bananas have few pests in Georgia.

CHERRY

Two major species of cherry are grown for fruit: sour cherry(Prunus cerasus) and sweet cherry (Prunus avium). Climate, specifically temperature, determines the success of cherry growing. In general, cherry does poorly in areas where summers are long and hot and winter temperatures are high for short periods. As a result, cherry does not do well in the south except at higher elevations. Cherry trees bloom very early in the spring (sweet cherries earlier than sour cherries) and are susceptible to damage by spring frost, so they are not reliable producers of fruit.

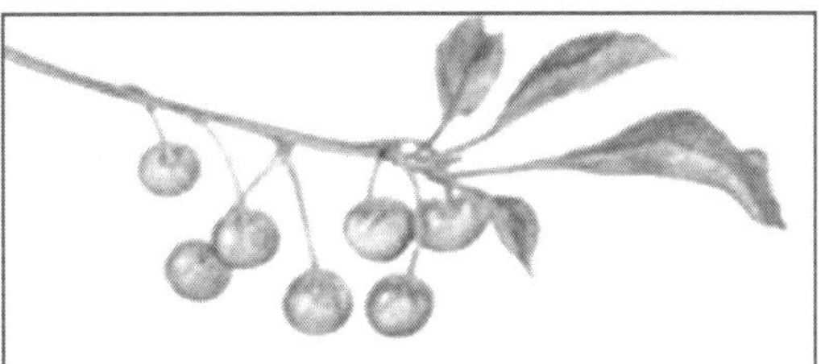

Cherries are similar to other stone fruits such as plums and peaches in respect to site requirements. Cherry grow on a wide range of soil types, so long as the soil is well-drained. Avoid soils that are extensively heavy or areas that remain wet for long periods. Plant cherry on a site that has good elevation and air drain-age to reduce changes of spring frost.

Information on variety selection is limited. Trials at Athens, Georgia, show fair success with 'Early Richmond' (sour), 'Montmorency' (sour) and 'Ranier' (sweet). In Virginia, trials show success with the sweet cherry varieties 'Hedelfinger,' 'Viva,' 'Valera,' 'Venus' and 'Hardy Giant.'

Sour cherry varieties are self-fertile (producing fruit with its own pollen) and are limited to areas such as the Piedmont and Mountains that receive 1,000 to 2,000 hours of winter chilling. The sweet cherries, with the exception of 'Stella,' are self-infertile (at least two varieties must be planted together to ensure cross-pollination) and are also limited to the Piedmont and Mountain sections of Georgia.

Cherry trees are typically trained to a central leader system similar to that used for apples and pears. This produces a central trunk with a lateral scaffold extending from the trunk. Pruning of young trees should be very light until trees come into bearing. Remove dead and broken limbs at first notice. Like peach and plum trees, cherry trees are very susceptible to an array of diseases and insects. Cherry leaf spot, brown rot and numerous viruses are the most serious diseases affecting cherry trees. Birds are another common problem and may consume an entire crop. Fruit splitting or cracking caused by rain on a maturing crop can cause losses of 75 to 80 percent.

CHINESE MULBERRY (CHE)

Chinese Mulberry (Cudrania tricus pidata) is also known as che, cudrang and silkworm tree. It is a small, thorny tree up to 30 feet tall. It flowers in late spring and the fruit ripens in the fall. The fruit has a raspberry-like appearance and a mulberry flavour. Propagation is usually by root suckers. Variety information is very limited. Che appears to be free of pest problems in Georgia.

ELDERBERRY

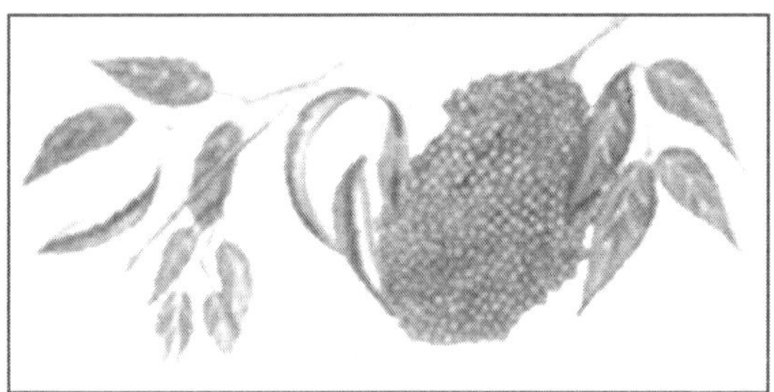

The American elderberry (Sambucus canadensis) is a large shrub that produces white flowers in the spring and large clusters of small, black fruit in late summer. The tree is drought hardy, winter hardy and attractive during blooming and fruiting. The small berries may be eaten raw when fully ripe or made into jelly, pies or wine. The berries should be picked when they are fully ripe and the colour is dark purple-black, but well before the time the berries begin to wither.

Green berries and stems are toxic and must not be eaten. Plant elderberries on well-drained or damp soil that has been limed to a pH of 6.0 to 6.5. Plant the bushes at least 5 feet apart and scatter a small amount of fertilizer (1/2 pound of 10-10-10) evenly under each bush every spring. Elderberries need limited pruning to remove dead and/or broken canes. They are easily propagated in the dormant season by transplanting sucker plants growing around the base of the other bush. Available varieties of elderberries include 'Adams,' 'Johns,' 'Nova,' 'New York 21' and 'York.'

FEIJOA

The feijoa or pineapple guava (Feijoa sellowiana) is a small, evergreen tree with attractive flowers and whitish-backed leaves. The plant is hardy to about 14 degrees F, so it is adapted to south Georgia. Severe freezes will kill the plant to the soil line, but it will regrow rapidly the following summer. Feijoas bloom in late spring and ripen in the fall. The fruit is egg-shaped and has a green skin with a yellowish pulp. The fruit has a pineapple/spearmint flavour, and the flowers have fleshy edible petals. The petals are sweet, and many people enjoy them as much as the fruit. In some countries, birds eat the petals and help with cross-pollination.

Adjust the soil pH to 6.0 to 6.5 prior to planting, and plant the bushes 8 to 12 feet part. Little is known about the fertilizer requirements of feijoa, but 1 to 2 ounces of 10-10-10 in March and July of the first year is suggested. Increase the amount by 1 ounce per application per year until the bushes are 8 years old; then use that amount for each application.

Water as needed to keep the bushes growing. Prune feijoas into small, single-stemmed trees. As the trees mature, remove selected branches to keep the trees somewhat open and productive. Keep the top of the tree narrower than the bottom to increase light penetration.

Available varieties of feijoa include 'Mammoth,' 'Triumph,' 'Coolidge,' 'Apollo,' 'Gemini,' 'Nazemetz' and 'Trask.' Most feijoas are partly self-pollinating but produce better crops if planted with three varieties for cross-pollination.

In Georgia, feijoas appear to have problems with transfer of pollen, since the blossoms are visited by few insects and birds. Hand pollination by snapping off a recently opened blossom of one variety and dusting it on the recently opened blossom of another variety may prove beneficial to increase fruit size and set.

GOOSEBERRY AND CURRANT

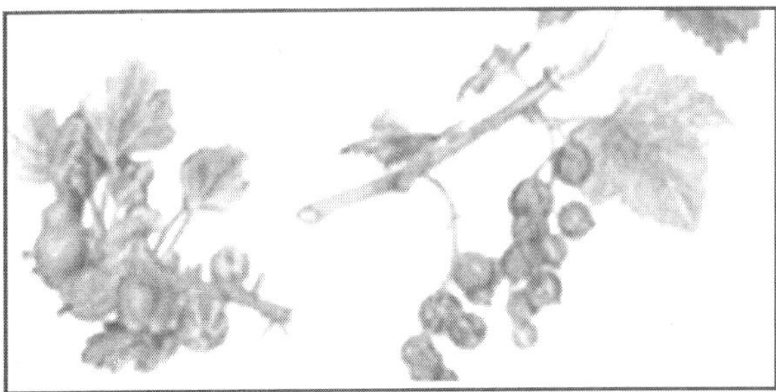

Gooseberries (Ribes Uva-crispa and R. hirtellum) and currants(Ribes sativum and R. nugrum) form small, beautiful arching bushes similar to blackberries in appearance. The crops can be used for making pies and jelly and for fresh eating. They are relatively pest-free but are not heat tolerant, so they are best adapted to north Georgia. They are very winter hardy.

Adjust soil pH to 6.0 to 6.5 prior to planting. Fertilize twice a season with 2 ounces of 10-10-10 for each foot of plant height; keep well watered. These plants are marginally adapted to the south. Common gooseberry varieties are 'Clark,' 'Fredonia' and 'Pixwell.' Common currant varieties are 'Red Lake,' 'Stephens No. 9,' 'White Grape' and 'Wilder.' Gooseberries and currants are intermediate hosts of white pine blister rust fungus. During the Depression, WPA crews destroyed wild and domestic gooseberries and currants in an effort to stem the disease. In recent years, it has been concluded that these plants are not a significant factor in the spread of the fungus.

JUJUBE

The Chinese jujube or Chinese date (Zizyphus jujuba) is adaptable to many areas in Georgia. It is drought hardy and produces a graceful, open-growing ornamental tree up to 40 feet tall. The plum-sized fruit has a thin, edible skin surrounding white flesh. The single hard stone contains two seeds. The immature fruit is green; mahogany-colored spots appear on the skin as the fruit ripens, and the fully mature fruit is entirely brown. When the fruit is brownish-green, it is similar to a mealy apple in texture and flavour. However, when fully dehydrated in a shriveled brown condition, the fruit is much better tasting and very similar to a date. The trees are quite cold hardy and bloom late enough in the spring that freezes seldom affect fruit production. The trees are self-fertile, but you can increase yields by cross-pollination with another variety.

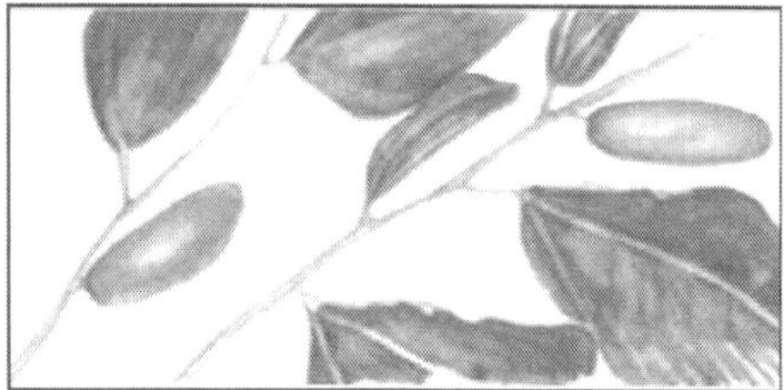

Jujubes can tolerate a very poor soil if it is well drained. Adjust the soil pH to 5.5 to 6.5 prior to planting. For best growth, watering may be needed during droughts. Fertilize lightly during the spring and summer with a balanced fertilizer such as 10-10-10. Pruning is seldom needed on jujubes, because the trees have an open, upright growth habit. Jujubes are propagated by budding or grafting onto a thorny rootstock. Jujubes usually produce numerous thorny root suckers that are useful for propagation but may become pests in some situations. Jujubes can be propagated from softwood cuttings.

More than 400 varieties of jujubes are known in China, but two of the best varieties in Georgia are the 'Li' and 'Tigertooth.' Jujubes have few pests in Georgia. Avoid spraying jujubes with dormant oil sprays used to kill scale insects on most fruit plants. Damage has been noted from dormant oil application in Florida.

JUNEBERRY

Juneberry (Amelanchier) is the common name for more than 25 species of small fruit-bearing trees or shrubs also called Serviceberry, Pacific Serviceberry, Saskatoon, sugar pear, Running Serviceberry and Allegheny Serviceberry. Most domesticated varieties have developed from Saskatoon Serviceberry (A.

alnifolia). Juneberry is widely adapted and extremely winter hardy. It is native to states from Maine to Iowa and south to northern Florida and Louisiana. Down Serviceberry is native to the Piedmont woods of Georgia. Other species are found as far north as Canada and as far west as the Great Plains. Prairie Indians mixed the fruit with buffalo meat and fat to make pemmican, their major winter food staple.

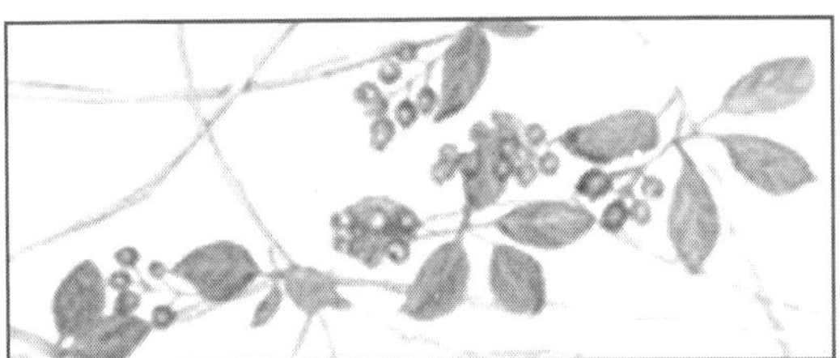

Juneberry is valued as an ornamental shrub and is commonly available. It forms multi-stemmed large shrubs or small trees that are usually 15 to 25 feet tall; they may grow to 40 feet under ideal conditions. The plant suckers form a bush-type growth if suckers remain. Small white flowers appear in mid- to late April during leaf development. In the fall, the foliage is highly colored, ranging from yellow to orange to rusty red. Fruit are berry-like pomes, ranging in size from that of a pea to 1/3 inch in diameter. The fruit of the Downy Serviceberry (A. arborea) is round; it is green, changing to red and to purplish-black upon maturity. The fruit ripens in June and is slightly sweet and juicy, with a mild flavour that has been compared to a combination of blueberry and cranberry. Birds are very fond of the berries.

Juneberry seems to thrive on a wide range of soil types, but it prefers a moist, well-drained acid soil. It will tolerate full sun or partial shade. It is susceptible to various pests including rust, powdery mildews and fruit rot as well as leaf miner, borers, pear leaf blister mite, pear slug sawfly and willow scruffy scale. Propagation is through softwood cuttings, but it is not very successful. Cold stratification of seed for 90 to 120 days at 41 degrees F is required. Division of parent plant and root cuttings are also used. 'Shannon,' 'Indian,' 'Success,' 'Regent,' 'Ovalis,' 'Grandiflora,' 'Prince Charles' and 'Jennybelle' are improved varieties of Juneberry.

KIWIFRUIT

No minor fruit has received more attention in recent years than the kiwifruit or Chinese gooseberry. It has been planted as an experimental commercial crop in many southeastern states and is available through some garden catalogs. There are two types of kiwifruit, the most common of which is the grocery store or commercial type (Actindia chinensis or A. deliciosa). The fruit grows on a vigourous vine with large, nearly round leaves the size of a saucer. The fruit is the size of a hen's egg and is brown on the outside and covered with fuzz.

The pulp is green and white with black seeds. The fruit has an acid flavour reminiscent of strawberries and watermelon. The vines are extremely cold-sensitive when young and may be damaged or killed to the ground by early fall freezes or late spring freezes. In midwinter, the vines are about as cold hardy as figs, withstanding temperatures to 10 degrees F.

The second type of kiwifruit is cold hardy enough to be grown in New England. Several species will grow in Georgia, including Actindia arguta, A. kolomikta and A. polygama, but there are few reports of heavy fruit production in Georgia. Most of the named varieties are derived from the A. argutaspecies. The fruit also grows on a vine, but these leaves are pointed and smaller than those of commercial kiwifruit.

The fruit is usually green, smaller than commercial kiwifruit and fuzzless. Fruits may be eaten like seedless grapes. 'Hayward' is the major commercial variety of female kiwifruit and has fruited fairly well in Georgia. The 'Bruno' variety has also performed well in Georgia.

Several varieties of male flowered commercial kiwifruit are available, but 'Matua' is probably the best in Georgia. Several varieties of cold hardy kiwifruit are available. 'Meader' and 'Anna' (Ananasnaya) are female varieties. 'Issai' is a self-fertile variety from Japan. Male vines may enhance the fruit set of 'Issai,' and the pollen from 'Issai' or male vines is needed for fruit set of the female varieties.

Like muscadine and bunch grapes, kiwifruit produce flowers on current season's growth that sprouts from last year's buds. Male and female vines of commercial kiwifruit must be planted to produce fruit. Usually one male is planted for every eight female vines. There are a few varieties of self-fertile cold hardy kiwifruit, such as the 'Issai' variety, but male vines are usually needed for cold hardy kiwifruit production. Kiwifruit require careful attention to water management. Irrigation is a must in growing kiwifruit to keep the vines from dying the first year.

They are the most drought sensitive fruit grown in Georgia, but they are also one of the most sensitive to overwatering. Kiwifruit grow best on a soil such as a sandy loam or sandy clay loam with good internal drainage. Raised beds are suggested in areas with marginal soil drainage at any time of the year. Adjust soil pH to 6.0 to 6.5 before planting.

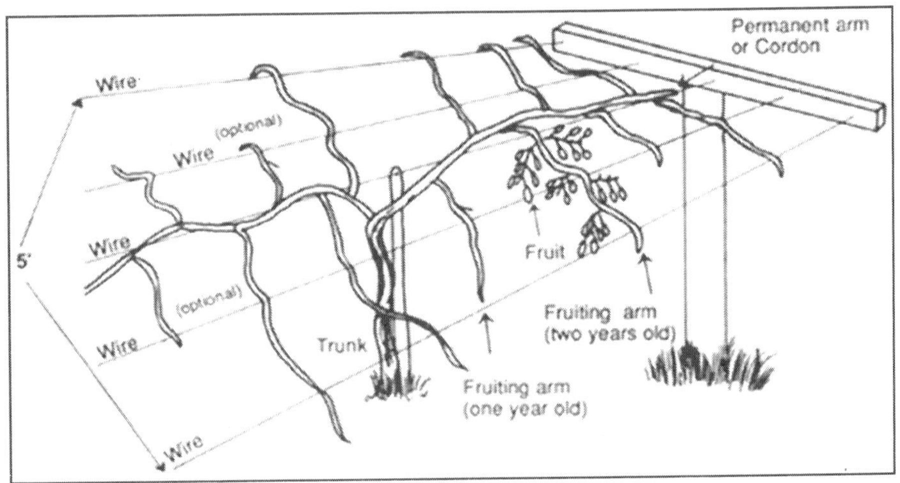

Fig. A Horizontal T-Bar Trellis for Kiwifruit.

Fertilize kiwifruit with 4 ounces of 10-10-10 in March, May and July of the first year. Scatter the fertilizer over a circle 24 inches in diameter around the plant. Increase this amount to 8 ounces the second year and to 1 pound the third year if the plants are growing well. Increase to 2 pounds per application for plants 4 or more years old if they have filled the trellis. Increase the area of fertilizer distribution as the plant grows. Kiwifruit need a strong trellis and require a significant amount of pruning. They may be grown on an overhead arbor (pergola) or on a T-bar trellis.

Set plants 8 to 15 feet apart depending on the amount of space available. The trellises should be 15 to 20 feet apart. In training a kiwifruit vine on a T-bar trellis, grow the vine as a single trunk to 6 inches below the wire. Then pinch out the top bud and train one shoot in each direction down the center wire to form a permanent arm or cordon.

Kiwifruit have a habit of growing vigourously for several feet and then going into a twining phase. It is best to prune off this growth and allow the next stage of vigourous growth to occur down the wire. Wrap the vine loosely on the center wire as it grows and tie it to the wire with degradable string, tape or cloth.

Allow fruiting arms to develop on both sides every 10 to 14 inches for commercial kiwifruit and every 24 to 30 inches for cold hardy kiwifruit. Allow fruiting arms to grow over the edge of the trellis and, if desired, to trail nearly to the ground. In the following year, the buds on these fruiting arms emerge and fruit is borne on the current season's growth. The next winter, remove the old fruiting arm if a replacement arm has grown. If no replacement arm is available, save the old arm and cut off last year's side shoots at 6 to 8 inches. New kiwifruit growth is very subject to wind damage, so tie new canes to the trellis as soon as possible. Kiwifruit can be propagated from cuttings in summer or winter is the right techniques are used. They may also be grown from seed, and budded or grafted to improved varieties. Cuttings are the best choice for commercial kiwifruit to reduce the danger of killing cold. The regrowth will always be true to variety for a cutting.

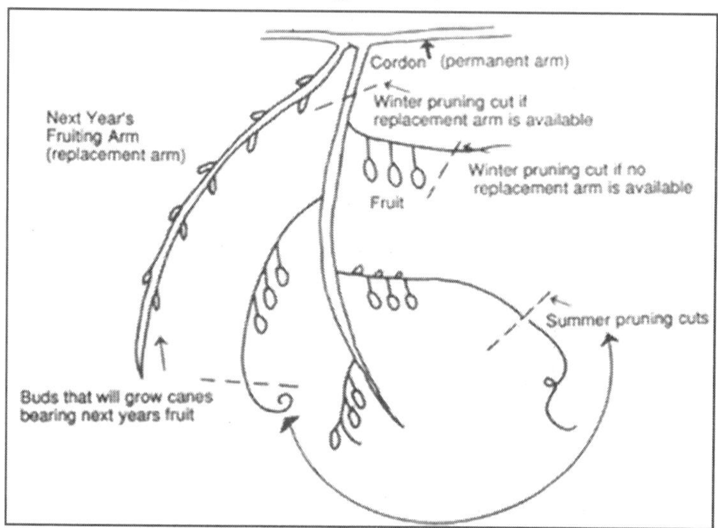

Fig. Pruning of Kiwifruit Fruiting Arms.

Kiwifruit have a number of pest problems. Root-knot nematodes are widespread in Georgia and are very destructive to kiwifruit. The home grower should sample his or her soil and plant kiwifruit in an area free of root-knot nematodes. Soil fumigation usually increases growth of kiwifruit and is recommended for the serious grower who is capable of handling the poison gas used in the fumigation process.

Root rot of kiwifruit is a serious problem when soils are too wet. Mulching is advised; however, do not use peanut hulls or peanut straw. Peanut litter can carry a root disease that is very destructive to kiwifruit. Several problem-causing insects have been observed on kiwifruit. Alder flea beetles (tiny, metallic greenish-blue beetles), cucumber beetles and grasshoppers may feed on the leaves. White peach scale and other scales are know to attack the plants. Deer and rabbits like to feed on the plants, and cats may damage some species of young kiwifruit.

LOQUAT

Loquat or Japanese plum (Eriobotrya japonica) is a handsome evergreen tree with a compact, rounded crown. Attractive white flowers are borne at the ends of the stems in the fall and winter, and fruit ripens in the spring. In Georgia, the new flowers and fruit are often destroyed by freezes if temperatures fall below 27 degrees F. Loquats are self-fruitful. The white, yellow or orange fruit are sweet with a mild plum-like flavour.

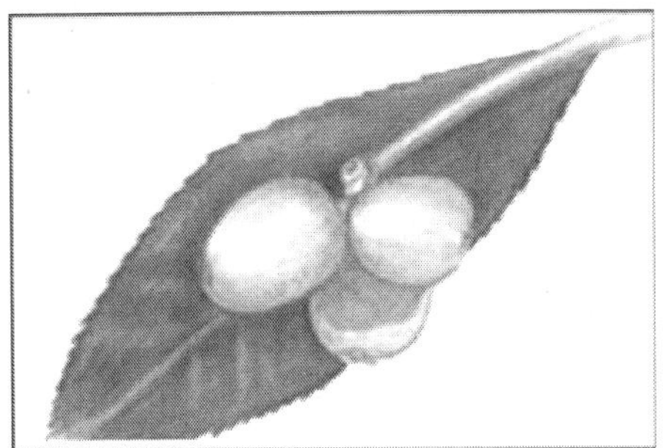

Mature loquat trees can survive temperatures as low as 12 degrees F but, because of their early blooming habit, they seldom fruit in north Georgia unless they are in a very protected location. In south Georgia, fruit production occurs about every three years in slightly protected locations. Plant loquats in the most protected location you can find with sun for at least half the day. Good choices are on the south side of a heated building and under tall pines.

Loquats are drought-tolerant but respond well to irrigation. They will not tolerate continuously wet ground. Adjust the soil pH to 5.5 or 6.5 prior to planting. Fertilize the trees lightly two or three times during the summer with a balanced fertilizer. Little pruning is needed except to remove dead and crossing limbs. Loquats are propagated by budding or grafting improved varieties onto seedling rootstocks. Some varieties that perform well in the southeast are 'Advance,' 'Bartow,' 'Fletcher Red,' 'Hardee,' 'Champagne,' 'Thales' and 'Wolfe.' Loquats have few pest problems. Occasional fire blight can usually be controlled by the prompt removal and burning of diseased parts.

MAYHAW

Mayhaws (Crataegus aestivalis, C. rufula or C. opaca) are in the rose family and hawthorne genus. They produce small apple-like fruits that ripen during late April and early May in south Georgia. The trees usually grow in swampy areas and may have their root systems flooded for several months each year. The fruits are highly prized for jelly. About 10 companies produce the jelly commercially, and it is available in many specialty stores in the southeast. Mayhaw trees are small- to medium-sized trees with beautiful white blooms and attractive fruit; they are desirable as ornamentals and for wildlife cover and forage. Some varieties have a low chilling requirement and bloom early (late February to early March in south Georgia), so these varieties should be planted only in south Georgia. Although mayhaws are naturally found in low, wet areas, they will grow well on hilltops with irrigation. Planting on hilltops will reduce the chance of freeze damage to the blooms during a radiation (still night) freeze.

Mayhaws are tolerant of wet soils but grow best in moist, well-drained soil. Adjust soil pH to 6.0 to 6.5 prior to planting, using dolomitic limestone. Fertilize the trees with 1 pound of 10-10-10 per inch of trunk diameter in early spring, up to a maximum of 5 pounds per tree. Repeat in July if the trees look pale. First-year trees should receive 1/4 pound of 10-10-10 in March and 1/4 pound in May and July.

Broadcast fertilizer evenly under the tree to avoid burning the roots. Do not apply fertilizer within 8 inches of the trunk. Mayhaw trees are long-lived and may have a 30-foot diameter canopy after 17 years, so plant trees 15 to 20 feet apart in the row with 18 to 20 feet between rows. If mechanical harvesting is desired, adjust row spacing to fit the equipment.

Train mayhaw to a single trunk at the base. The first branches should start at 24-36 inches so equipment can be operated under the tree or people can get under the tree and pick up the fruit. Occasional pruning is necessary to open up the tree for greater light penetration. The trees will adapt to a modified central leader training system when one main trunk is promoted by pruning. This is a common method of training apple trees.

Mayhaws can be grown from seed collected from ripe fruit and sprouted immediately without cold treatment. A more dependable method is to clean the seed and store it in damp sand at 33 degrees F for eight months prior to planting. Plant seed in sterilized soil and cover with a screen to prevent birds from eating the young seedlings.

The seed of hawthorns is unusual in that many of the seedlings will be true to type. Mayhaw seed viability varies greatly among trees. Mayhaw cuttings can be rooted under intermittent mist or in a humidity chamber during the summer. Dipping in a root-promoting hormone (2,000 to 6,000 ppm IBA) may help rooting. Propagation from hardwood and root cuttings has been reported. Plants grown from cuttings will produce fruit like the plant from which the cuttings were taken.

Mayhaws are easily grafted in late winter. In south Georgia, a whip and tongue or simple whip graft can be used in early to mid-February. Cleft grafting can be used on larger trees. Mayhaws appear to be initially graft-compatible with any hawthorn, but using mayhaw rootstock is the safest bet. In Mississippi, the parsley haw is considered an excellent rootstock. The hoghaw, which grows on Georgia's sand ridges, can be used as a rootstock, but its slow growth rate may allow the mayhaw scions to grow over the hoghaw rootstock.

Mayhaws grafted onto 'Washington' hawthorn are severely dwarfed and have performed poorly. Mayhaw seedlings are probably the best choice as rootstock, especially in damp soils. About 70 mayhaw trees have been selected from the wilds of Georgia, Mississippi, Louisiana and Texas and named as varieties. Information and observations are very limited on some varieties. Most ripen over a 20-day harvest period, but the 'Lori' variety may have 80 percent of the fruit ripe at one time. Early blooming, low chilling varieties face a significant danger from spring freezes, but they provide very early ripening fruit in south Georgia in years when they are not frozen out. In middle and north Georgia, plant only the highest chilling, latest blooming mayhaw varieties available such as 'Saline,' 'Crimson' and 'Texas Star' to reduce the change of spring freezes.

Insects, including plum curculio, aphids, flat-headed apple borers, white flies and foliage feeders are known to attack mayhaws. Plum curculio has caused extensive damage to fruit in some locations and requires a spray programme in most areas. Deer and rabbits can destroy a grove in a short period of time. There are numerous diseases known to occur on various hawthorn species. Information on diseases of mayhaw is limited. Quince rust is a common fruit disease of mayhaws. Cedar trees are an alternate host for the disease, and removing all cedar trees in the vicinity or planting mayhaws away from existing cedars should help. Brown rot blossom blight can attack the developing shoots in the spring during some years. Fire blight occurs on mayhaw but is usually not a severe problem.

Table. Some Promising Mayhaw Cultivars Tested in South Georgia in Approximate Order of Ripening.

Cultivar	Typ. Date Ripe	Typ. Bloom Date	Fruit Size (mm)	Fruit Color (Skin/Flesh)	Firm	Retention	Comments
T.O. Superberry	late Apr	mid Feb	15-18	dk. red/pink	8	7	early bloom a problem
Mason's Superberry	late Apr	mid Feb	17-18	dk. red/pink	8	8	early bloom a problem, holds on tree tightly
Superspur	late Apr/ early May	late Feb/ early Mar	17-18	lt. red-yellow/ yellow	5	5	productive, poor color, soft, shatters
Saline	late Apr/ early May	early-mid Mar	15-19	mostly red/lt. pink	8	8	late blooming, productive, good firmness & retention
Big Red	late Apr/ early May	early Mar	18-19	red/pink	7	7	good size and color
Crimson	late Apr/ early May	mid Mar	16-18	mostly red/lt. pink	6	5.5	late blooming
Big V	late Apr/ early May	late Feb/ early Mar	16-18	lt. red/ pinkish-yellow	5	5	most of crop ripens in early May, soft, shatters
Turnage 57	early-mid May	early-mid Mar	13-16	red/yellow	5	5	late ripening, late blooming, soft, shatters
Texas Star	early-mid May	early-mid Mar	19-22	red/yellow	6	8	promising but not extensively tested yet
Turnage 88	mid May	early-mid Mar	15-17	red/yellow	5.5	6	late ripening, late blooming, somewhat soft

Because few pesticides are labeled for mayhaw, commercial producers may have trouble controlling pests. Promote maximum tree health through proper site selection, irrigation, fertilization and weed control to prevent many of the diseases that might occur in the woody parts of the tree.

Ratings are on a 1-10 scale where 7 is very good and 8 excellent, except in the case of fruit retention where 8 is slightly excessive.

MEDLAR

The medlar (Mesphilus germanica) is a small shrub-like tree in the rose family. It is cold hardy as far north as New York. Large, white blossoms appear in the spring and develop into small russet brown fruit, which are hard and acidic. The fruit are harvested after a light frost and stored where they are allowed to mellow, at which time they become edible.

Medlars can easily be grown from well-matured seed, or they may be grafted on pear or quince rootstock. Common varieties include 'Hollandish,' 'Nottingham' and 'Dutch Royal.'

MULBERRY

Mulberries are large, fast-growing trees that are good fruit producers for humans and wildlife. The fruit resembles a slender blackberry and have a mild flavour. The fruit drops when ripe and may be harvested by shaking the tree. The fruit are borne on current season's growth and ripen in May in Georgia. The three species cultivated are the red mulberry (Morus rubrum), native to the United States; the black mulberry (Morus nigra), native to Iran; and the white mulberry (Morus alba),native to Japan and China. There are usually few problems with plant hardiness in Georgia's climate; the plants, however, can be damaged by late spring freezes. Mulberries grow best on a deep, well-drained soil, but they are tolerant of poorer soils. Adjust the soil pH to 5.5 to 6.5 prior to planting.

Fertilize the trees in late winter and in mid-summer, using about 1 pound of 10-10-10 for each inch of trunk diameter. Prune trees each winter to remove dead and crossing branches. Mulberries are easily propagated from hardwood cuttings in late winter. Many mulberry varieties are available, including White Mulberry ('Downing,' 'New American' and 'Wellington'), Black Mulberry ('Black Persian' and 'Noir'), and Red Mulberry ('Hicks,' 'Johnson,' 'Stubbs,' 'Townsend' and 'Travis').

Mulberry trees are relatively pest-free except for white peach scale, which attacks twigs and trunks of the trees and may kill them. White peach scale can be controlled by several dormant oil sprays in late winter. Birds are extremely fond of ripe mulberries and compete with humans for the fruit. Some white mulberries occasionally suffer from popcorn disease. The disease is caused by a fungus that causes individual fruit carpels to swell until they look like unpopped popcorn kernels. Collect and destroy diseased fruit.

PAWPAW

The pawpaw (Asimina triloba) is the only cold hardy species of the "custard apple" family. The paw-paw is a native American fruit with northern growing limits of New York, Michigan and Ontario, Canada. The range extends south to Florida and west to Nebraska and Texas. Tree pawpaw forms a small tree with a short trunk and spreading branches, forming a rounded crown. Tree height and width at maturity is 15 to 20 feet, and trees will grow to 30 to 40 feet under ideal conditions. Trees tend to send up root suckers, resulting in the formation of a thicket or grove of trees.

The dwarf pawpaw (Asimina parviflora) is a low growing shrub, seldom more than 6 feet tall. The flag pawpaw (Asimina incana) is another type of dwarf pawpaw. It has attractive white flowers in the spring, but the fruit is small. Tree pawpaw leaves have a medium green upper surface and a lighter green lower surface. Leaves tend to droop, giving the tree a sleepy appearance.

Flowers are inconspicuous, maroon to purple in colour and 1 to 2 inches in diameter. Pawpaws bloom in early May, just as leaves are developing. Fruit are borne in clusters of one to six, depending on the success of pollination. Fruit size and shape vary greatly. Fruit is from 2 to 6 inches long and is elongated or rounded. Fruit contain numerous medium- to dark-brown seeds resembling elongated lima beans. Seed size varies from pea size to 1 1/4 inches long.

Fruit have a very thin green skin, which turns yellowish-black when ripe, like an overripe banana. Fruit ripen from September until frost. After ripening, fruit soften and perish rapidly. The flesh has a rich, sweet custard consistency and a strong nutty banana flavour. Fruit are high in food value, with more than 430 calories per pound.

Pawpaws are relatively free of disease and insects compared to most cultivated fruits. They do best on fertile, well-drained soils that are slightly acid, and they grow well in full sun or dense shade. Transplanting is difficult and should be done when trees are small (3 to 6 inches tall). The transplants should be balled and burlapped and should receive special care to keep stress such as drought and weeds at a minimum during establishment.

Trees can be propagated by seed or by layering and root cuttings. Suckers are difficult to transplant. Seed should be stratified in a moist medium for 60 days at 41 degrees F before being sown in the spring. Germination may be erratic. Seeds can also be sown outside in the fall to germinate the following spring. Many different varieties are available including 'Davis,' 'Mango,' 'Mitchell,' 'P A Golden,' 'Sun-flower,' 'Taylor,' 'Taytow,' 'Wells,' 'Wilson' and 'Overleese.' In the early 1990s, there was a pawpaw variety trial at Tifton, Georgia. 'Overleese' was the best performing variety in this trial. The 'Mango' variety was selected from Georgia.

POMEGRANATE

Pomegranates (Punica granatum) are dense, bushy shrubs 6 to 12 feet tall with thorny, slender branches that may be trained into small trees. Orange-red flowers appear on new growth in the spring and summer and are bell-shaped and vase-shaped. The vase-shaped flowers are normally sterile, so they will not develop into fruit. Pomegranates generally fruit poorly in Georgia. The fruit contains numerous seeds surrounded by sweet pink, juicy, subacid pulp covered with leathery-brown to red, bitter skin, which is easily peeled. Pomegranate juice stains can be difficult to remove from clothing. Pomegranates may be damaged by unseasonably low temperatures in the fall, winter or spring and in mid-winter by temperatures below 10 degrees F. Pomegranates can tolerate many soil types and some flooding. Pomegranates grow best on a deep, fairly heavy, moist soil at a pH range of 5.5 to 7.0. Proper watering is important in growing pomegranates because adequate soil moisture is necessary to control fruit splitting and reduce fruit drop. Fertilize young pomegranates with 1 pound of 10-10-10 in March and July. Increase the rate as the plants grow until the mature tree is receiving 3 pounds of 10-10-10 in March and July.

Most growers prefer to train pomegranates into a multiple-trunk system. Select five or six vigourous suckers and allow them to grow. Pomegranates require some pruning each year, and unneeded vigourous shoots should be removed. The short spurs on two- or three-year-old wood growing mostly on the outer edge of the tree produce flowers. Light annual pruning encourages growth of new fruit spurs. Heavy pruning reduces yield, so be careful to leave adequate fruit-bearing wood on the tree while removing branches that may cross over or interfere with growth.

Hardwood cuttings are usually used for propagation. Cuttings 8 to 10 inches long of wood ¼ to ½ inch in diameter are cut in winter from the previous season's growth and planted with 2 to 3 inches of the top exposed. Several varieties are available, including 'Belgal,' 'Granada' and 'Early Foothill' (early ripening), 'Ruby Red,' 'Sweet Spanish Papershell' and 'Wonderful.' However, most of these varieties only set a few fruit each year in Georgia. In north Florida, 'Belgal' has been more productive than other varieties and usually produces about 10 fruit per year. There are many door-yard trees of unknown varieties around old home places and plantations that set good crops of fruit most years. These can be propagated by hardwood cuttings. Pomegranate leaf blotch or fruit spot are occasionally problems.

QUINCE

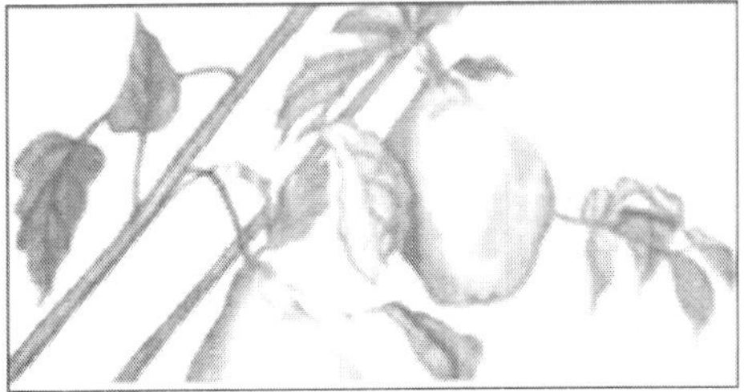

Several types of quinces are grown in Georgia. The common quince (Cydonia oblonga) forms a small tree with attractive flowers and leaves. Flowers form on the end of new growth in the spring and usually are self-fertile. Fruits are round or pear-shaped and weight up to 1 pound. They are very hard and are edible when cooked or made into jelly. A second species of quince, common or Japanese flowering quince (Chaenomeles speciosa), is grown as an ornamental for its red blossoms, which appear early in the spring. This bush produces oblong, yellow fruit that makes good jellies and jams. Quinces can survive neglect and are tolerant of a wide range of soils. Fertilize lightly to discourage vigourous growth that may succumb to fire blight. Train common quince trees to a vase shape. Drooping branches may need to be shortened. Little pruning is needed on mature trees except for removing water-suckers. Available varieties include 'Champion,' 'Apple,' 'Pineapple' and 'Smyrna.'

Quince is subject to many of the diseases that attack apples and pears, including fire blight, but it is generally much less affected. Quince rust is an exception, and some quince varieties may suffer crop losses in heavy rust years. Fire blight can be a problem, particularly if trees are excessively vigourous.

ALMOND

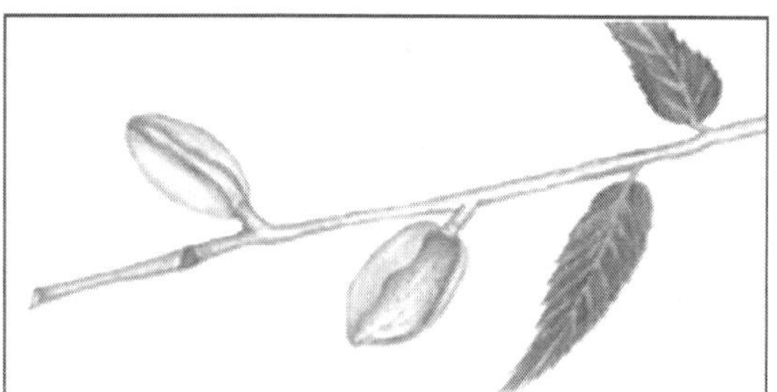

The almond (Prunus amygdalus) is a close relative of the peach. The tree and snow white blooms are similar to those of the peach, but the seed is the edible part of the almond. Climate requirements are quite exact for almonds, and commercial production is limited to areas with dry summers. The necessary chilling hours (300 to 500 below 45 degrees F) are much like the peach varieties grown in south Georgia, but rain and high humidity during the growing season (late July and August) cause nut rot and inhibit nut opening. The outer flesh of the almond must have dry weather to dry and split open properly. Most almond varieties require cross-pollination, but some self-fertile varieties are available. Almond culture is the same as peach culture, and the recommendations on planting, fertilizing and pruning peaches are applicable to almonds. For Georgia, the best home orchard variety is 'Halls Hardy.' This variety is late blooming, self-fertile and hardshelled. The same pests are a problem for almonds and peaches, and a regular spray schedule is required to produce a good yield.

CHESTNUT

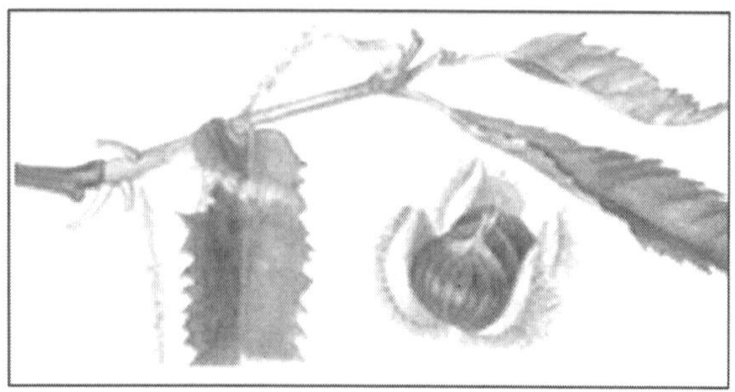

The major chestnut species found in Georgia are the American chestnut (Castanea dentata) and Chinese chestnut (Castanea mollissima). Minor chestnut species such as Japanese, Korean, European or Italian, and chinkapins are occasionally seen. The chestnut blight fungus Endothia parasitica was introduced from Asia in the late 1800s. It spread through the east, and by the early 1950s, virtually all American chestnuts had been destroyed. The blight does not kill the root system, so new sprouts grow and form small trees that are reinfected and killed. The blight resistance of the Chinese chestnut was recognized in the early 1900s, and most plantings since then have been of the Chinese variety. Chinese chestnut is as hardy as the peach and is adapted to most of Georgia. Chestnut is monoecious, with separate male and female flowers on the same tree. The burr (fruit) of the true chestnuts (American, Chinese, European and Japanese) normally contains three chestnuts. The chinkapins (Castanea pumila) have one nut per burr.

Chestnuts grow best in well-drained soil. Adjust the pH to 5.5 to 6.0 prior to planting. Fertilizer and care recommendations are the same as for most temperate climate nuts, and spacing of 20 feet is ideal for chestnuts. Chestnuts must be harvested every other day; numerous fungi and bacteria attack the nuts on the ground, causing rapid decay and spoilage. Use gloves when harvesting; the burrs can be painful to handle.

Propagation of chestnuts is by nursery graft and inlay bark graft. The varieties of Chinese chestnuts recommended for home planting are 'Crane,' 'Nanking' and 'Meiling.' Pests affecting chestnuts are chestnut blight, chestnut weevil and gall wasp.

BLACK WALNUT

The black walnut (Juglans nigra) is a dual value tree because of its timber and nuts. Timber provides the greater income potential from black walnut. The trees are massive, and some grow 98 feet high and more than 7 feet in diameter. Black walnut is native to North America and does well in Georgia in well-drained soils. Black walnut is monoecious, like other walnuts, pecans and chestnuts. Black walnuts grow best on deep, well-drained, moist and fertile soils with a pH of 6.0 to 7.0. Establish black walnuts by transplanting seedlings from nursery beds or by planting nuts directly in the field. Spacing depends on whether the primary crop is nuts or timber, but 30 feet by 30 feet is acceptable, followed by later thinning. The trees should be pruned to a central leader to produce a straight, long, single trunk because of the high timber value. Nut production is generally secondary. Of the many varieties available, 'Thomas' is one of the best for nut production. Two varieties — 'Victoria' and 'Captain' — are reported to be resistant to anthracnose, which is a fungal disease that attacks trees.

CARPATHIAN WALNUT

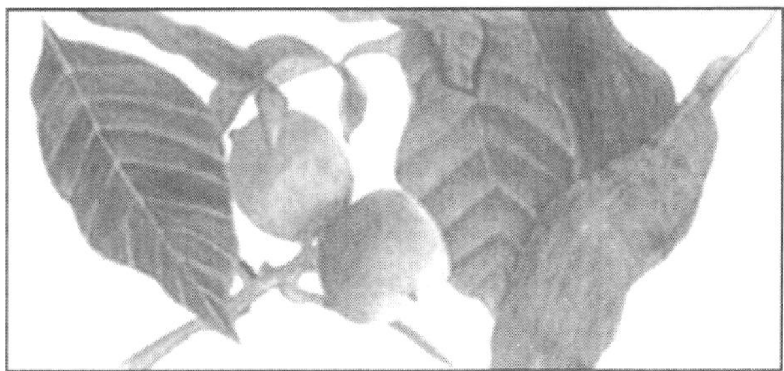

The Carpathian walnut (Juglan regia), also known as Persian or English walnut, is believed to have originated near the Caspian Sea in Iran. The Greeks and Romans brought it to the west, and it was eventually spread throughout the Far East and North America. Carpathian walnuts are sensitive to climate extremes and are marginally adapted to Georgia. They are quite susceptible to spring frost injury. Temperatures below 12 degrees F will kill some varieties, and temperatures above 100 degrees F will cause sunburn on limbs and shrivel kernels. The walnut is monoecious, but cross-pollination is required; so at least two varieties should be planted. Carpathian walnuts do best on deep, well-drained soils with a pH of 6.0 to 7.0. A spacing of 25 feet by 25 feet is ideal. Carpathian walnuts are propagated like pecans, but no specific variety recommendations can be made.

NEW CULTIVARS OF ORNAMENTAL FIELD PLANTS

Plants of this group have been grown as minor cut flowers, but recent introduction of new cultivars, with modified and improved horticultural traits, turned them into important floral crops. Examples are:

ANIGOZANTHOS HYB., HAEMODORACEAE (KANGAROO PAW)

This Australian plant was grown mainly outdoors until a few years ago. Recently introduced highly yielding interspecific hybrids are now grown indoors for year round production. These new hybrids are propagated by in vitro tissue culture.

ASTER HYB., ASTERACEAE

New interspecific hybrids of A. novi-belgii and other species native to Eastern North America turned these herbaceous perennial, late summer garden plant, into an important greenhouse crop. This is a LD-SD plant, requiring at first LD, until the stems reach a certain desired height and then it is exposed to natural winter SD.

CAMPANULA MEDIUM L., CAMPANULACEAE

This plant, native to south Europe, was used only as garden and pot-plant until recently. The original species required a long cold period followed by LD for flowering . However, new varieties, introduced recently, have long flowering stems and require only LD for flower induction. This enables growing the plant as a commercial cut flower crop.

CLARKIA AMOENA NELS. AND MACBR., ONAGRACEAE (GODETIA, SATIN FLOWER)

This Western North American plant was mainly a garden plant until the recent introduction of improved cultivars for use as cut flowers. The plant requires mild temperature and moderate watering and feeding. It is a facultative LD plant.

EUSTOMA GRANDIFLORUM SHINN (SYN. LISIANTHUS RUSSELLIENUS), GENTIANACEAE

Native to Southern US, it was used sparsely as garden and cut flower plant. Newly introduced F_1 hybrids turned the plant into an important greenhouse cut flower crop for year round production. Seed propagated, it requires mild-low temperatures in the first growing stage, followed by warmer temperatures.

LEUCADENDRON HYB., PROTEACEAE

This South African shrub became an important outdoor crop for cut flowering shoots with the introduction of new hybrid cultivars. The 'Safari Sunset' cultivar is now grown on over 200 hectares in Israel.

LIMONIUM HYB., PLUMFAGINACEAE

Interspecific hybrid cultivars of several perennial limoniums became important greenhouse cut flower crop, used as "filler."

GARDEN AND LANDSCAPING PLANTS

These are mainly woody or herbaceous perennials, used for many years in gardens and introduced recently into the floral trade. Examples are:

COTINUS COGGYGRIA SCOP., ANACARDIACEAE (SMOKE TREE)

A deciduous shrub, native to South Europe, used for many years as a garden plant. The cultivar 'Royal Purple' is now grown for cut foliage. LD is applied to prevent plants from entering dormancy.

HYPERICUM SP. (HYPERICACEAE)

Several species and hybrids of these shrubby plants, native to the Mediterranean and the Canary Islands, have recently became important floral crop grown both outdoors and in greenhouses for cut shoots with fruits of various colours. This is an absolute LD plant that requires night lighting for winter production.

RUSCUS HYPOGLOSSUM L. (LILIACEAE)

This herbaceous perennial has been grown in Israel as a garden plants for many years. It is now the main cut foliage crop in Israel, grown exclusively in shaded houses.

ORNAMENTAL CULTIVARS OF FIELD CROPS

In some plants, grown mainly as field crops, new ornamental cultivars have been introduced and used as cut flowers. Examples are: sunflower (Helianthus annuus L., Asteraceae), cotton (Gossypium hirsutum L., Malvaceae), and safflower (Carthamus tinctorius L., Asteraceae).

PLANTS GROWN IN BOTANICAL GARDENS

Botanical gardens and specialized plant collections are rich sources for plant material, some of which can be used for introduction as potential floral crops. Some examples are the bulbous plants of the Liliaceae: Eremurus sp. of Central Asia, the South African Bulbinella kookerri of yellow, orange, and white flowers, and Ornithogalum dubium of yellow and orange flowers, and the South Asian Curcuma alismatifolia (Zingiberaceae).

WILD PLANTS IN THEIR NATIVE HABITAT

The introduction and development of Geraldton wax-flower described above is an example of such introduction. Some such plants are currently under intensive developmental stages. They include plants originated from remote areas, but also plants native to Israel and California.

2

Cut Flowers

Cut flowers, whether purchased from the florist or cut from your own garden, will last much longer in the vase if you follow the simple guidelines below.

USE A CLEAN VASE

Start with vases that have been cleaned with hot soapy water to eliminate bacteria and fungi and then rinsed thoroughly.

PREPARE THE FLOWERS PROPERLY

Cut just-opening flowers early in the morning and place in water immediately. The vase life of flowers that ooze a milky "sap" (poinsettias, poppies) may be improved by immersing the bottom 2 inches of their stems in boiling water for 10 seconds before using them in an arrangement. Gently remove lower leaves from the stem so there will be none in the vase water. Before you put any flowers in the vase, recut the stems, removing 1 to 2 inches at an angle under water. You can do this in a basin full of water, or even by holding the stem and the blades of the shears (or kitchen scissors) under running tap water. Don't crush or burn flower stems. In our experience these practices are of little value.

WATER

Flowers in most arrangements collapse early because they are unable to obtain enough water to keep them looking crisp and fresh.

There are a number of ways to ensure that your flowers get enough water:

1. Recut them under water to ensure that no air gets into the stems.
2. If you live in a hard water area (you find white deposits in teakettles and on faucets), use demineralized water sold in supermarkets for filling steam irons, to make your vase solutions.
3. Use a vase solution which is hot but not uncomfortable (100 degree F).
4. Use one of the following suggested vase "preservatives."

Never use softened water in a vase solution as it contains sodium, which is bad for cut flowers.

FOOD

Flowers are living things, and like us they need food for proper growth and healthy colour. Amazingly, you can provide much of what a cut flower needs with one of the following simple vase solutions. They contain acid to improve water flow in flower stems, sugar to help buds open and last longer, and a preservative to reduce growth of bacteria and fungi.

1. Mix one part of any of the common lemon-lime sodas with three parts of water. Do not use diet drinks or colas. Diet drinks have no sugar and the colas contain too much acid for flowers. Adding 1/4 teaspoon of household bleach (Chlorox or similar) per quart will keep the solution clear.
2. Put 2 tablespoons of lemon juice or bottled "Real Lemon," 1 tablespoon of sugar, and 1/4 teaspoon of bleach in a quart of warm water. Add another 1/4 teaspoon of bleach to the vase every 4 days.
3. Use a commercial flower preservative. These are sold in florist shops and supermarkets but may not be as effective as the above recipes for improving flower vase life. However, they are inexpensive and very convenient to use; simply follow the directions on the packet.

Don't use aspirin or vinegar in vase solutions; they are rarely effective in increasing vase life of flowers.

USE ARRANGING AIDS PROPERLY

If you are using florist foam as an arranging aid, let it soak in the vase solution until it sinks. Do not push it down into the container as air bubbles will remain inside the foam and cause early flower death. Insert stems carefully.

KEEP FLOWERS COOL

The higher the temperature, the faster flowers will deteriorate, so it is advisable to cut them early in the morning, when temperatures are cool, and to avoid exposure to heat. Don't place arrangements in sunny locations, near heaters or fireplaces, or on top of television sets. Do put arrangements in a cool place overnight if you possibly can.

USES OFCUT FLOWERS

Cut flowers are flowers or flower buds (often with some stem and leaf) that have been cut from the plant bearing it. It is usually removed from the plant for indoor decorative use. Typical uses are in vase displays, wreaths and garlands. Many gardeners harvest their own cut flowers from domestic gardens, but there is a significant commercial market and supply industry for cut flowers in most countries.

Fig. Rose, Hydrangea, Calla Wedding Bouquet.

Fig. Flower Garland Sellers Outside Banke Bihari Temple, Vrindavan, India.

Fig. A Flower Market in Vietnam.

The plants cropped vary by climate, culture and the level of wealth locally. Often the plants are raised specifically for the purpose, in field or glasshouse growing conditions. Cut flowers can also be harvested from the wild. The cultivation and practices of raising cut flowers form a part of horticulture. They are often included in that branch of horticulture called floriculture.

USES

A common use is for floristry, usually for decoration inside a house or building. Typically the cut flowers are placed in a vase. A number of similar types of decorations are used, especially in larger buildings and at events such as weddings. These are often decorated with additional foliage. In some cultures, a major use of cut flowers is for worship; this can be seen especially in south and southeast Asia.

Sometimes the flowers are picked rather than cut, without any significant leaf or stem. Such flowers may be used for wearing in hair, or in a button-hole. Masses of flowers may be used for sprinkling, in a similar way to confetti. Garlands (especially in south Asia), and wreaths (in Europe and the Americas) are major derived and value added products.

LONGEVITY OF CUT FLOWERS

Live cut flowers have a limited life. The majority of cut flowers can be expected to last several days with proper care. This generally requires standing them in water in shade. They can be treated in various ways to increase their life. In most countries, cut flowers are a local crop; because of their limited life after harvest they have to be marketed quickly. In India, much of the product has a shelf life of only a day. Among these are marigold flowers for garlands and temples, which are typically harvested before dawn, and discarded after use the same day. There is also a market for 'everlasting' or dried flowers, which include species such as Helichrysum bracteatum. These can have a very long shelf life.

COMMERCE

The largest producers are, in order of cultivated area, China, India, and the United States. The largest importer and exporter by value is the Netherlands, which is both a grower and a redistributor of crops imported from other countries. Most of its exports go to its European neighbours.

In recent decades, with the increasing use of air freight, it has become economic for high value crops to be grown far from their point of sale; the market is usually in industrialised countries. Typical of these is the production of roses in Ecuador and Colombia, mainly for the US market, and production in Kenya and Uganda for the European market. Some countries specialise in especially high value products, such as orchids from Singapore and Thailand. The total market value in most countries is considerable. It has been estimated at approximately GBP 2 billion in the United Kingdom, of the same order as that of music sales.

KEEPING CUT FLOWERS AND FLOWERING PLANTS

CUT FLOWERS

Whether cut flowers are grown in a home garden or in a greenhouse by commercial experts, their care is a science.

To keep cut flowers beautiful longer; remember that they have been removed from their source of water, the root system, and will wilt quickly if not placed in water. Cut stems should be placed in water immediately, as air will rapidly move into the water-conducting tissues and plug the cells. This is why the cut flower that has been out of water more than a few minutes should have a small portion of the lower stem cut off so that water will move up freely when it is returned to water. Cuts can be made under-water to assure no air enters the stem. A cut flower also has been removed from a major source of food—the leaves on the plant to which it was attached. Although the leaves on the flowering stem make food, once indoors they are in a reduced light situation and this limits available carbohydrates.

Use a Preservative

Commercial preservatives will increase the life of cut flowers and should always be used. (Adding aspirin, wine, or pennies to cut flowers WILL NOT help to keep them fresh longer. Do not attempt a home brew concoction.) A floral preservative is a complex mixture of sucrose (sugar); acidifier, an inhibitor of microorganisms; and a respiratory inhibitor. Sucrose serves as a source of energy to make up for the loss of the functioning leaves and insures continued development and longevity of the flower.

An acidifier makes the pH of the water more near the acid pH of the cell sap. Most water supplies are alkaline and can reduce the life of cut flowers. The acidifier also stabilizes the pigment and the colour of the flower. This is why red roses turn "blue" when placed in water without a preservative or acidifier. A microorganism growth inhibitor is perhaps the most important part of a floral preservative. Bacteria and fungi are everywhere and are ready to enter the cut surface of the stem and multiply. Prior to actual decay symptoms, cells of the water-transporting tissues can become blocked with microorganisms, inhibiting water uptake.

To aid the floral preservative in slowing down microorganisms, always clean the vase or container. Also remove all leaves below the water surface, as they soon deteriorate. Water and water uptake are major factors in keeping cut flowers fresh.A process called "hardening" ensures maximum water uptake. It simply means placing the freshly cut stem in 110° F (43.5° C) water (plus preservative). Place in a cool location for an hour or two. Maximum water uptake is attained because water molecules move rapidly at 110° F (kinetic energy) and quickly move up the stems. Flowers at cool temperatures lose less water. In this one brief period while the water is cooling, freshly harvested stems, leaves, and flowers take up almost as much water as in the balance of their life.

Other Tips for Long-Lasting Cut Flowers

Check the water level of the container or vase daily and add water plus preservative when needed. Keep flowers away from hot or cold air drafts and hot spots (radiators, direct heat, or television sets).

While both drafts and hot spots increase water loss, hot spots reduce a flower's life by speeding transpiration (water loss) and respiration (use of stored food such as sugars) and increasing development (rate of petal unfolding). When away from home, move the flowers into the refrigerator or the coldest (above 35° F/1.5° C) spot in the house. Again, this will slow down water loss, respiration, and development.

Never store fruit and flowers together. Apples produce ethylene gas, a hormone that causes senescence, or aging, in flowers.

In summary, to keep cut flowers longer:

- Recut the stems and remove excess foliage.
- Harden the flowers by setting them in warm water in a cool place.
- Use a floral preservative.
- Keep them cool and avoid drafts, hot spots, and television sets.
- Use a clean vase or container and check the water level daily.

FLOWERING PLANTS

Inadequate light, high temperatures, and improper watering are the common causes of failure in flowering plants. These plants are grown in a greenhouse where the night temperature is usually cool, the air is moist, and light is ample. When these plants are brought into a dry home where the light is poor and the temperature is maintained for human comfort without consideration for the plants, the results are often disappointing.

Poinsettias

Poinsettias require bright light and should be kept away from drafts. A temperature between 65° and 70° F is ideal. Avoid temperatures below 60° and above 75° F. Keep the plants well watered but do not over-water. Newer, long-lasting varieties can be kept attractive all winter.

Reflowering a Poinsettia

The poinsettia blooms during short days. Starting October 1, exclude poinsettia from artificial light for 16 hours; either cover with a light-proof box each evening or place in an unlighted room or closet. Expose to full light during the day (eight hour days). Use fertilizer when new growth is visible. After 10 weeks of short days, the plants should reflower.

Easter Lilies

Fig. Keep the Plants in a Sunny Place where the Temperature does Not Fall Below 60° F. Water when the Soil Feels Dry.

After the plants have turned brown, cut off the stem at the soil surface. When the garden soil warms up in late May, move the plants outdoors. Choose a warm sunny place with well-drained soil. Plant the bulbs four to six inches deep (soil surface to the top of the bulb) in most soils and somewhat deeper in sandy soil. Easter lilies may bloom the first fall after being set outdoors. They are easy to transplant.

Fertilize several times during the summer and use a mulch to keep the soil moist. In the fall when the soil is lightly frozen, apply evergreen boughs or marsh hay around the plants. Keep this mulch on until new growth develops the next spring.

AZALEAS

Azaleas require direct sunlight to remain healthy. A night temperature of 50° to 60° F will prolong blooming. Keep the soil moist. Azaleas can be planted in a shady spot in the garden during the summer months. Feed them with an acid fertilizer and examine them frequently, keeping plants watered during dry periods. Greenhouse azaleas will not survive Zone 4 winters.

Azaleas need short days and cool temperatures to form flower buds. If you can provide short days, after buds have formed, a six- to eight-week cool treatment is needed before plants will bloom. A well-lighted room with a temperature of 35° to 50° F is ideal, but hard to find in most homes. Unless you have the proper growing conditions for the azalea, you should not attempt to carry the plants over to the next year.

CYCLAMENS

Cyclamens require full sunlight and a night temperature of between 50° and 60° F. Flower buds will fail to develop if the night temperature is too high or if the light is poor. They require regular watering. Plants can be carried over, but as with the poinsettias, homegrown plants are seldom equal to those grown by a commercial grower.

GARDENIAS

Gardenias grown indoors need special care and specific conditions. They demand an acid soil and should receive the same nutritional care as azaleas. The night temperature should be near 60° F and the humidity around the plant should be kept high. High temperature and low light intensity will result in flower-bud drop.

HYBRID ROSE (ROSA HYBRIDA)

VARIETIES

Gladiator, Baby Pink, Sofia Lawrence, YCD 1, YCD 2, YCD 3 are commonly cultivated.

Fig. YCD 1.

Fig. YCD 2.

Fig. YCD 3.

SOIL AND CLIMATE

It is generally suitable for higher elevation (1500 m and above). It can also be grown in the plains under ideal condition of fertile loamy soils with salt-free irrigation water. The ideal climate for rose growing should have temperature with a minimum of 15°C and maximum of 28°C. Light is important factor which decides the growth. The growth is slowed by day length, *i.e.* > 12 hours and heavy overcast, cloudy/mist conditions. High relative humidity exposes the plant to serious fungal diseases. In tropics the ideal temperature is 25°C – 30°C on sunny day and on cloudy day 18°C – 20°C. The optimum temperature should be 15°C – 18°C. These temperatures are extremely difficult to find and it's therefore to compromise.

PROPAGATION AND PLANTING

The crop can be propagated by rooted cuttings or by budding on Briar root stocks in hills and on Edward Rose and Rosa indica in plains. One year old budded plants are planted in July - August at 75 cm × 75 cm spacing.

PLANTING OF ROSE

After Cultivation

The plants should be watered daily until they establish and thereafter once in a week. Pruning is done during March and October. Spray Diuran 2.5 kg a.i/ ha to control weeds. Avoid spray fluid coming in contact with Rose plants.

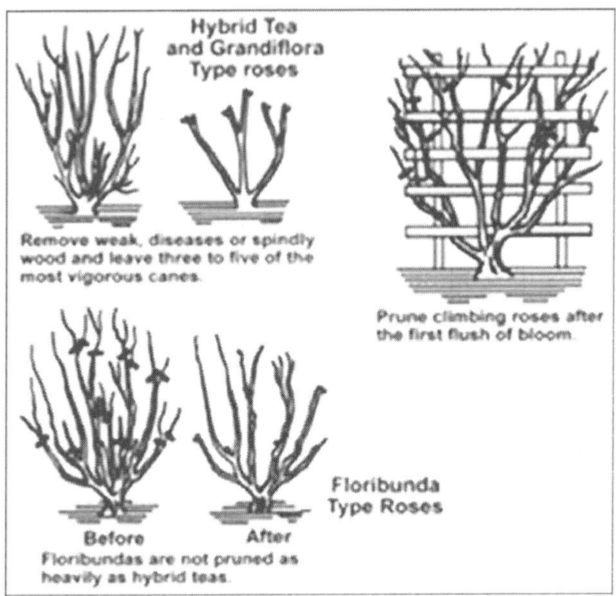

Fig. Pruning.

SUPPORT OF THE PLANTS

Post is placed at internals of 3m on both sides of the bed. Along the sides of the bed, galvanized wires or plastic string are fastened at the posts at 30cm – 40cm intervals to support the plant. Between the wires across the bed, thin strings can be tied to keep the width of the beds constant.

DISBUDDING

Varieties produce some side buds below the center bud. These side buds have to be removed or disbudded. The disbudding must be done regularly and also as soon as possible in order to avoid large wounds in the upper leaf axil.

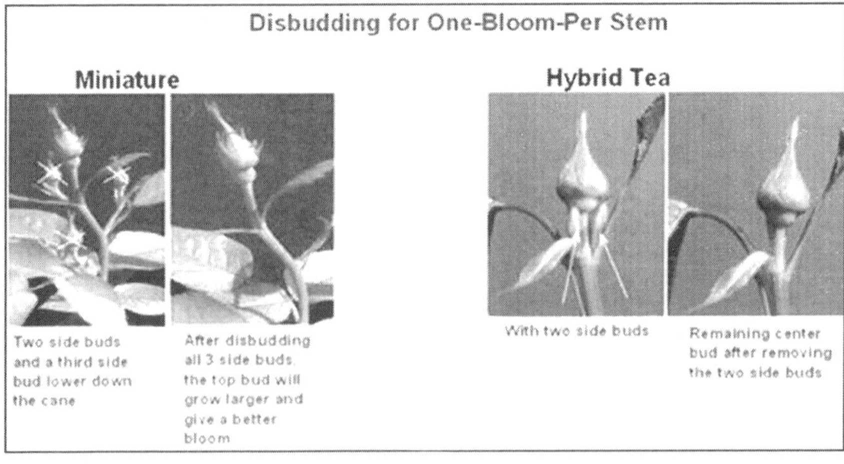

DEAD SHOOT REMOVAL

In the old plants the dead shoot or dried shoots on plants will serve as the host for fungi. So regularly these have to be removed.

SOIL LOOSENING ON BEDS

After 6 months or so, there is every chance that the soil become stony and it has to be loosened for efficient irrigation.

BENDING

Leaf is a source of food for every plant. There should be balance between Source (Assimilation) and sink (Dissimilation). After planting, 2 to 3 eye buds will sprout on main branch. These sprouts will grow as branches and these branches in turn form buds. The mother shoot is bend on 2nd leaf or nearer to the crown region. The first bottom break or ground shoot will start coming from the base. These ground shoots form the basic framework for production and thereon the ground shoots should be cut at 5th five pair of leaves and medium ground shoots should be cut at 2nd or 3rd five pair of leaves.

Fig. Bending.

DEFOLIATION

The removal of leaves is known as defoliation. It is done mainly to induce certain plant species to flower or to reduce transpiration loss during periods of stress. Defoliation may be done by removal of leaves manually or by withholding water. The shoots are defoliated after pruning.

MANURING

At three months interval, apply FYM at 10 kg and 8:8:16 g NPK/plant after each pruning. For cv. Happiness NPK may be applied @ 75:150:50 g/plant/year.

HARVEST

Harvesting is done with sharp secateure at the tight bud stage when the colour is fully developed and the petals have not yet started unfolding. There should be 1-2 mature leaves (those with five leaflets) left on the plant after the flower has been cut. The reason for leaving these matures leaves is to encourage production of new strong shoots. Harvesting is done preferably during early morning hours.

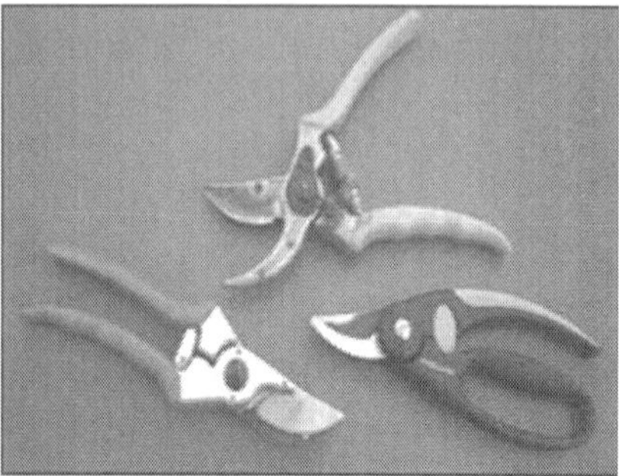

Fig. Secateur for Harvest.

Fig. Harvesting Technique.

POSTHARVEST HANDLING

Roses must be placed in a bucket of water inside the polyhouse immediately after harvesting and transported to cold storage (2-4°C). The length of time depends upon the variety and quality of the roses.The flowers are graded according to the length. It varies from 40-70 cm depending on the variety and packed in 10/12 per bunch.

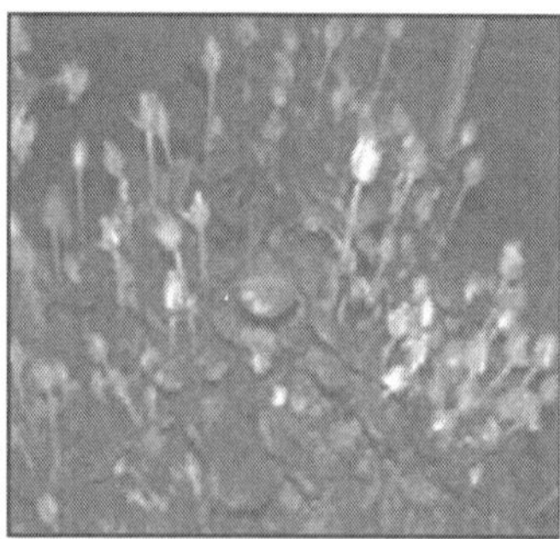

Fig. Pre Cooling.

Fig. Grading.

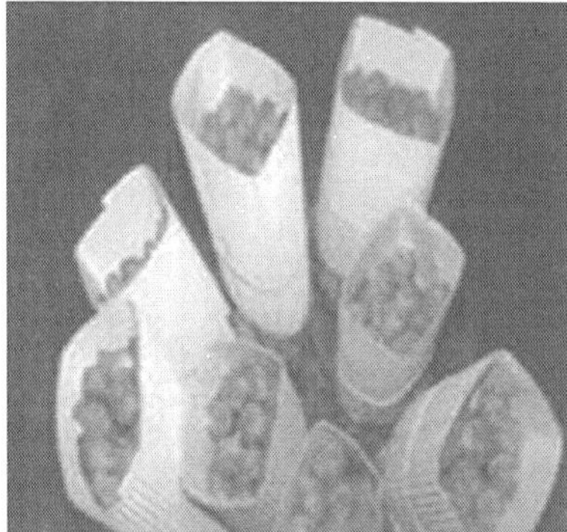

Fig. Packing.

PHYSIOLOGICAL DISORDERS

Blind Wood

The normal flowering shoot on a greenhouse rose possesses fully expanded sepals, petals, and reproductive parts. The failure to develop a flower on the apical end of the stem is a common occurrence. Such shoots are termed as blind wood. The sepals and petals are present, but the reproductive parts are absent or aborted. Blind wood is generally short and thin, but it may attain considerable length and thickness when it develops at the top of the plant. This may be caused by low temperature, insufficient light, chemical residues, insect, pests, fungal diseases and other factors.

Bull Heads or Malformed Flowers

The center petals of the bud remain only partly developed and the bud appears flat. They are common on very vigourous shoots, particularly bottom breaks, and it is possible that there is a lack of carbohydrates to develop the petals. The cause of bull heading is yet unknown, however, thrips infestation will also cause malformed flowers. Also at low temperature, some varieties will form bull heads.

Colour Fading

The off- coloured flowers are seem to be a problem with some yellow varieties. In these varieties the petals may be green or a dirty white instead of a clear yellow.

Raising the night temperature several degrees will reduce the number of off-coloured flowers. Occasionally the pink or red varieties develop bluish-coloured flowers.

This is very often associated with use of organic phosphate and various other kinds of insecticides.

Limp Necks

The area of the stem just below the flower "wilts" and will not support the head. This may be due to insufficient water absorption; cutting off the lower 1 to 2 inches of stem and placing the cut stem in water at 37°C will revive the flower.

Blackening of Rose Petals

This is caused by low temperature and high anthocyanin content. GA3 treatment causes accumulation of anthocyanin in petals of Baccara roses. This effect was more pronounced at low temperature (20°C at day and 4°C at night) than in higher temperature (30°C at day and 20°C at night).

Nutritional Disorders

Iron deficiencies can cause pale foliage. Adjusting the pH of the soil may solve this problem.

Yield

The Hybrid Teas roses can yield about 70 – 80 stems/plant/year, while the Floribundas yield yields 80 -90 stems/plant/year.

CARNATION (DIANTHUS SPP)

VARIETIES

Standard Types

Red - Domingo, Master, Gaudina, Leopardii, Big Red, Taureg, Guapo, Aicardii

White - Baltico, White Liberty, Emotion, White Dona, Lisa

Pink- Dona, Charmant, Dumas, Pink Dover, Bizet

Light pink - Charmant, Cipro Big Mama, Dona, Golem

Yellow - Diana, Kiro, Soto, Salamanca, Liberty

Orange - Solar, Star, Folgore

Double - Malaga, Star, Athena, Happy Golem

Fig. Pink Dona.

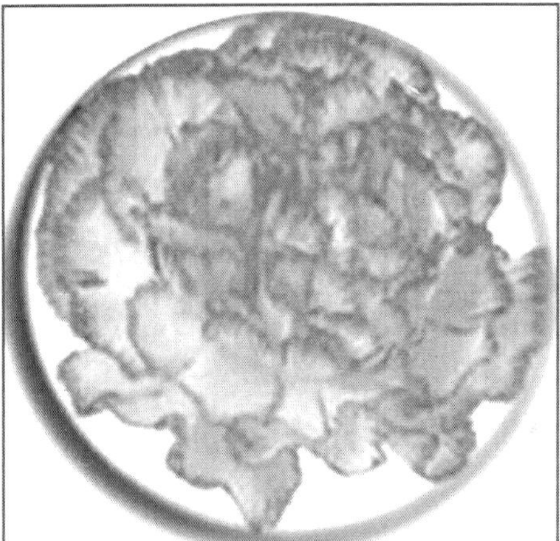

Fig. Tundra.

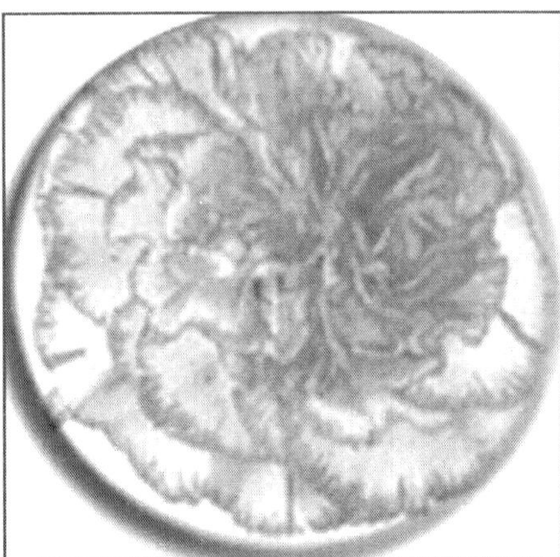

Fig. White Tundra.

SPRAY TYPES

Estimade, Indira, Vera, Durago, Amore, Kiss Siga are some of the popular spray types.Red- Red Eye, Red Fuego, Red Vital, AveiroWhite- White Prestige, Milky Way, Elvis,T-587Pink- Rosa Bebe, Spur, Suprema, D- 925, Celebration, OsirisYellow - Stella, Prestige, Mila, Sonia, AbrilOrange - Sunshine, Autumn, Fancy Fuego, Disney, EilatDouble - Berry, Orbit Plus, Nadeja, Picaro

Fig. Spray Types.

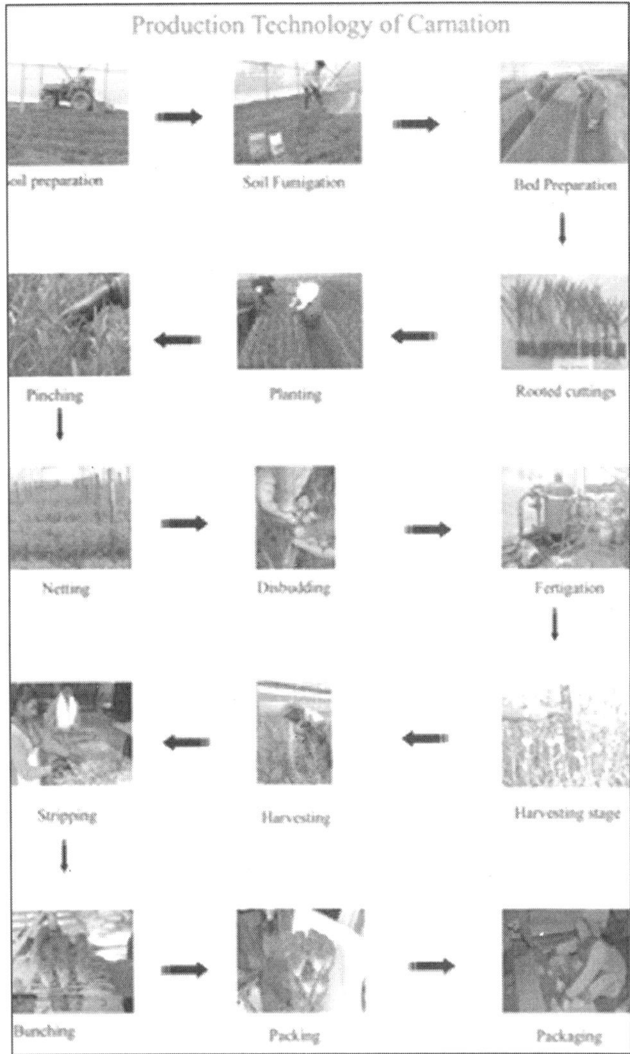

Growing environment: Naturally ventilated aerodynamic steel frame structure. Climate: Cool climate with day temperature of 18-24^{0}C and night temperature of 10-15^{0}C; relative humidity of 70 -75 per cent.

Soil: Well drained red loamy soil with pH of 5.5 - 6.5.

SEASON

It can be cultivated throughout the year as it is grown under controlled conditions.

PROPAGATION AND PLANTING

Plantlets/suckers can be used for planting. The terminal cuttings of 5-10 cm are treated with NAA at 500 ppm for 5 minutes to induce rooting. Cuttings are dipped in Carbendazim 2g/lit solution. Raised beds at 3 feet width and 45 cm height are formed at 45 cm interval and planting is done on top of the bed at 15 × 15 cm spacing. The cuttings normally develop good root system within 21 days. Fumigation - Dazomet @ 30g/m^2 or H_2O_2 @ 300 ml/m^2.

*Bed size:*100 cm width, 30 cm height, convenient length with 40 cm foot path.

Spacing: 6 row planting - 15×15 cm (25 plants/m^2). 4 row planting - 15×15cm (22 plants/m^2) .

Irrigation: Drip system with drippers at 30 cm spacing (5-6 l/m^2/day). Growing condition – Day temperature 20-25°C Night temperature 10-15°C Critical photoperiod 13 hours RH 50-60 per cent Planting method in carnation

Nutrition: The following fertigation schedule can be adopted for intensive production under polyhouse conditions.

Nutrients	Quantity (g/m2/week)	
	Till Bud Formation	Bud Formation to Harvest
Tank-A (Monday and Thursday)		
Ammonium Nitrate	3.0 g	2.0 g
19:19:19	3.0 g	2.0 g
Magnesium Sulphate	2.5 g	2.5 g
Boron	1.0 g	1.0 g
Trace elements / micronutrients	1.0 g	1.0 g
Tank – B (Tuesday and Friday)		
Potassium Nitrate	5.0 g	5.0 g
Calcium Nitrate	8.0 g	9.0 g

SPECIAL PRACTICES

Netting for plant support: 4 layers
1st layer : 7.5 × 7.5 cm
2nd layer : 10 × 10 cm

Fig. Netting Practice in Carnation.

3rd layer : 12.5 × 12.5 cm
4th layer : 15 × 15 cm

SUPPORT MATERIAL

Carnation crop has the tendency to bend unless supported properly. Hence the crop needs support while growing. Good support material is metallic wire woven with nylon mesh. At every two meters the wire should be supported with poles.

The poles at both the ends of bed should be strong. Metallic wire is tied around the bed along the length with the support from supporting poles. Across the bed, nylon wires are woven like net. For an optimum support, an increasing width of the meshes can be used. Bottom net can be of 10×10cm, then two nets of 12.5×12.5cm and the upper most can be 15×15cm.

Pinching:

- Depending upon the need of crop spread, single, one and a half or double pinch method is adopted.
- Ideal time for pinching is early morning.
- When the plant attains 5 nodes, the first pinch is given. This is called 'single pinch'. This would give rise to six lateral shoots.
- With a 'one and half pinch', 2-3 of these lateral shoots are pinched again. For the 'double pinch', all the lateral shoots are pinched off.

Fig. Pinching in Carnation.

DISBUDDING

In standard carnations, side buds should be removed whereas in spray carnations, the terminal bud has to be removed.

IRRIGATION

Irrigation is provided with drip system once in 2-3 days according to soil moisture to maintain water holding capacity at 60 per cent to 65 per cent. The optimum water requirement of the crop is 4-5 lit/m^2/day

MANURING

Neem cake 2.5 ton/ha, Phosphorus 400 g/100 sq.feet and Magnesium sulphate 0.5 kg/100 sq.feet are applied as basal.

TOP DRESSING

Calcium Ammonium Nitrate and MOP at 5:3 ratio is mixed and applied @ 2.5 g/plant/month as top dressing. Field overview in different stages of crop

Fig. Plant Protection.

APHIDS - MYZUS PERSICAE

Damage Symptom

Feeding usually occurs in buds and undersides of leaves. Feeding on young leaves results in distorted leaves as they continue to grow. Older leaves may display patches of chlorolic spots. Nymphs and adults suck the sap from the leaves, stems and flower buds in colonies.

Control

Spraying the plants with Thiomethoxam 1ml/litre or Imidachloprid @ 2ml/litre or Asatap (Acephate) @ 0.5 to 1.0 gm/litre.Thrips: Thrips tabaciDamage symptom Both the nymphs and adults suck the sap from leaves and flower. They excrete brown droplets, which afterwards true black. Leaves may fade and shrivel in case of heavy infestation and foliage becomes silvery.

Control Measures

Spraying of Fipronil 1.5 ml/lit (or) Imidachloprid 2ml/litre or Dimethoate 30 EC @ 1ml litre.

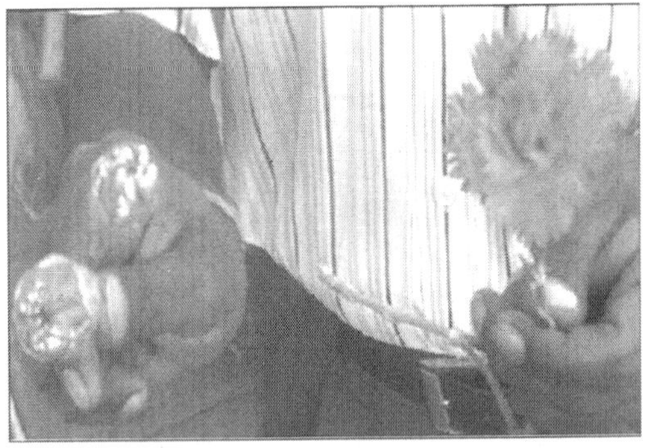

Fig. Thrips Infested.

RED SPIDER MITES-TETRANYCHUS URTICAE

Damage Symptom

These have ability to produce fine silk webbing, spider mites are very tiny and very small and are difficult to identify. They suck sap from the leaves which results in tiny yellow or white speckles. Once the foliage of a plant becomes bronze it often drops prematurely. Heavily infested plant may be discoloured stunted.

Fig. Red Spider Mite.

Fig. Infested Bud.

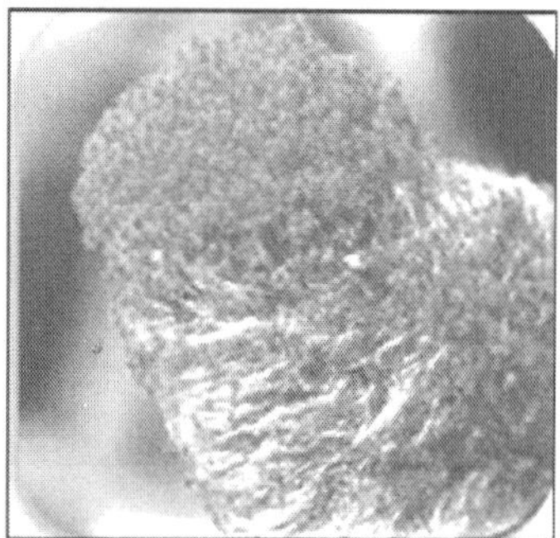

Fig. Ballooning Symptom.

Control :

- Apply Abamectin (Vermitec) 1.9EC @ 0.5ml/litre
- Spray Azardiractin 50,000ppm 3ml/litre
- Discard the plant and leaf debris.

Bud Borer – Helicoverpa Armigera.

Caterpillers infest the leaves and flower buds. Bore holes are clearly visible on flower buds. Finally the infested flower buds are fail to open. The attack by this pest is most during warm season.

Control :

- Spray Bacillus thuringiensis 2g/litre.
- Release 1 lakhs of Trichogramma egg parasitoid per acre.
- Set up Helilure sex pheromone trap @ 4 Nos/acre
- Spray spinosad 0.75ml/litre or Thiodicarp 0.5 ml/litre

Diseases

- *Fusarium wilt:* Soil drenching with Carbendazim @ 0.1 per cent or Difenoconazole @ 0.05 per centorPseudomonas fluorescens as soil application @ 25 g/m^2 and foliar application @ 0.5 per cent at monthly intervals orsoil drenching with Bacillus amyloliquefaciens @ 0.5 per cent at monthly intervals

- *Alternarialeaf spot:* Bacillus subtilis as soil application @ 25 g/m2 followed by foliar application @ 0.5 per cent at monthly intervals

PHYSIOLOGICAL DISORDER

- *Calyx splitting:* Spray borax @ 1 g/l at fortnightly intervals till flower bud appearance and at weekly intervals thereafter.
- *Harvest:* Flowering starts 110-120 days after planting.

Stages of Harvest

Standard types - Paint brush stage

Spray types - When two flowers are open and the remaining flower buds show colour Yield: 15 flowers/plant (350 - 375 flowers stems/m2) in 2 years period.

FUSARIUM WILT

Fig. Grey Mould Rot.

DISORDER

Calyx Splitting

Cultivars with too many petals are susceptible to calyx splitting. Varying temperature and environmental conditions also influences calyx splitting. The calyx may split down either half or completely. The petals are deprived of their support, which results in bending down of petals. Thus, the regularity of shape and structure of the flower get destroyed. Selection of cultivars like Epson, Palmir etc. that are less prone to splitting, regulation of temperature and maintenance of optimal fertilizer level can minimize this disorder. This can also be reduced by placing a rubber band or 6mm wide clear plastic tape is used around the calyx of the flowers which are just start opening.

Curly Tip

This disorder affects the growing tips which curl and become distorted. Tips of the young shoots fail to separate and continuation of growth results in a characteristic curvature. Poor light and other adverse conditions are thought to be the causes of the disorder. Water stress and potassium deficiency are suspected causes for a physiological curly tip and die-back of carnation flowers.

Season of Flowering Development and Harvest

Flower starts after 4 months of planting and continues up to one and half years. Standard carnation flowers are harvested when the outer petals unfold nearly perpendicular to the stem. Spray types are harvested when two flowers open and the remaining buds show colour. Daily harvest is made leaving bottom 5 nodes of stalk to facilitate side shoot development.

Post Harvest Treatment

Citric acid is added to water to make the pH 4.5 to 5 and 5 mg of Sodium hypochloride is added to 1 litre of water. Cut flower stalk is soaked in this solution for 4 – 5 hours to improve vase life. After harvest, the flower stems have to be trimmed at the base and should be immediately placed in a bucket of preservative solution of warm and deionized water. A good preservative solution for carnations should be acidic (pH 4.5) with 2-5 per cent sucrose and a biocide not phytotoxic to carnations. After keeping in preservative solution for 2 to 4 hours, flowers should be placed in a refrigerated room at 0-2°C for 12-24 hours. The flowers can be stored for two to four weeks before marketing. For this, the flowers have to be packed in cartons lined with polyethylene film. These cartons should have sufficient vent holes. The full cartons should be pre-cooled with out lid. The plastic is then loosely folded on top of the stems and the lid is closed. These cartons are stored in cool chambers designed to maintained 0°C with good air circulation and a constant relative humidity of 90-95 per cent.

Grading

Based on stem thickness, stem length and quality of flower grading is done as A, B, C, D.

Yield

The average yield is about 8 Stalks/plant/year

Fig. Grading and Packing in Carnation.

*Precision production techniques for carnation

S.No.	Cultural Practice	Recommendation
1.	Fumigation	Dazomet @ 30 g/m^2
2.	Media consortium	10:1:1 ratio of 30 kg/m2of consortium with 25 kg of Farm Yard Manure, 2.5 kg of vermicompost, 2.5 kg of cocopeat with the biofertilizers Azospirillum, Phosphobacteria, VAM and the biocontrol agents *Trichoderma viridae, Pseudomonas fluorescens* each @ 20 g/m^2 at bimonthly intervals
3.	Planting density	15 x 15 cm with 25 $plants/m^2$
4.	Planting stage and pinching level	30 day old rooted cuttings and single pinching at the 5th node
5.	Precooling	$4^{\circ}C$ for 4 hours
6.	Pulsing solution	Sucrose 10 % + Citric acid 100 ppm + 8-Hydroxy Quinoline 400 ppm for 24 hours duration
7.	Holding solution	Sucrose 5% + Citric acid 50 ppm + Benzyl Adenine 75 ppm
8.	Wrapping and packaging techniques	Polyethylene sleeves 50 gauge thickness + CFB with 4 % vent

* The precision production techniques are to be followed along with the cultural practices recommended for the conventional system.

ANTHURIUM (ANTHURIUM ANDREANUM)VARIETIES

- *Red :* Temptation, Tropical Red, Red Dragon, Verdun Red, Flame, Mauritius Red .
- *Orange :* Mauritius Orange, Peach, Casino, Sunshine Orange, Nitta.
- *White :* Acropolis, Linda de Mol, Mauritius White, Lima, Manoa Mist.
- *Pink :* Abe Pink, Candy Stripe, Passion.
- *Green :* Midori, Esmaralda.
- *Bicoloured :* Titicaca, Jewel, Akapana, Cardinal.
- *Others :* Fantasia (cream with pink veins), Chocos, Chicos (chocolate brownish red).

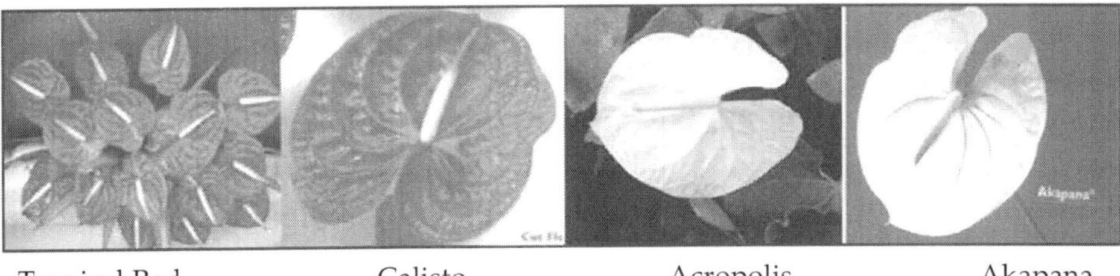

Tropical Red. Calisto. Acropolis Akapana

Flame Cheers Aymara Caesar

Meringue Coto Paxi Castano Cerilla

Tropic Night Condor Eesmeralda Fantasia

Grace Laguna Lima White Marshall

Midori Pistache Poopo Safari

Titicaca

ClimateAnthurium requires porous, well drained aerated soil rich in organic matter content. The soil pH should be 5.5 and 6.5. It performs well under green shade net having 70 – 80 per cent shade intention with 80 -90 per cent humidity and 24 - 28°C temperature and 15 - 22°C night temperature with 1500 – 2000 foot candles light intensity.

Fig. Under Shade Net.

Growing environment: 75 per cent shade net house with 70 - 80 per cent relative humidity, day temperature of 24 - 280C and night temperature of 15 - 220C.Growing mediaA growing media containing 1:1 mixture of leaf mould and coco peat with a pH of 5.5 to 6.5 is ideal, which ensures good drainage as well as water holding capacity.

Fig. Coco Peat Bricks.

PROPAGATION:

Propagated through tissue culture or suckers. Tissue culture plants are widely used for commercial cultivation. Seed : Seeds germinate within 10 days; transplanted after 4-6 month takes 2 - 3 years to bloom. Seeds scattered on a finely shredded medium and kept under 75 per cent shade.

Also germinated aseptically under nitsch/ms media supplemented with BAP and Adenine Suckers : Suckers produced from base of the plant at 4-5 leaf stage with 2-3 roots separated.

57 ppm BAP at monthly intervals on more than one year old plant encourage more suckersStem cutting : Top of the stem with few roots of 3 to 4 year old plants is removed and planted. Each cutting should have single eye or bud IBA 500ppm produce good roots.

Tissue culture : Becoming popular; explants – leaf segments, root segments, stem section, vegetaive buds, flower stalks, spathe and spadix; MS mediumPlanting: Grown in pots or raised beds. Tissue culture plants of 15 cm height with 4-6 leaves are ideal for planting.Irrigation: Mist or over head sprinkler to provide water and to improve relative humidity.

Stem cutting Suckers Tissue culture plants

Young plants under Pot culture Spike ready for harvest

POT CULTIVATION

Foliar application of 0.2 per cent of NPK @ 30:10:10 during vegetative stage and 10:20:20 during flowering stage is adopted for pot cultivation. Fertigation can be adopted for raised bed cultivation.

RAISED BED CULTIVATION

For the first 6 months spray plants with a solution of cow dung and DAP @ 250 ml/plant (10 kg of cow dung + 2 kg of DAP dissolved in 200 l of water and the decanted solution is used for spaying). After 6 months fertigation is adopted with the following schedule.

Fertilizer	Quantity (g/100m^2)
Schedule 'A' - Weekly once	
Calcium Nitrate	250
Potassium Nitrate	150
Micro nutrients	50
Schedule 'B' - Weekly once	
Mono Ammonium Phosphate	250
Potassium Nitrate	100
Magnesium Sulphate	50

BED SYSTEM

Soil is incorporated with organic matter. Bed size of 1.2 to 1.4m width with a spacing of 60 × 60 cm is found ideal.

SHADE REGULATION

Open condition with adequate shading facility are the best. Growing under polythene plastic with shade cloth prevents bacterial blight. 70-80 per cent shade level is found to be best for Tamil Nadu and Kerala conditions. Excess light causes permanent damage to the leaves. Shade net should be laid at a minimum height of 3m from ground level.

Fig. Under Polythene Plastic with Shade Cloth.

FERTILIZER REQUIREMENT

NPK @ 30:10:10 @ 0.2 per cent is given from 30 days of planting as foliar application at weekly intervals

GROWTH REGULATORS

Application of GA3 200 ppm as foliar spray at 2 month intervals improves the growth and quality of flowers

AFTER CULTIVATION

Leaf pruning retaining 4 – 6 leaves/plant has to be taken up then and there to avoid disease problem and to promote flowering. The roots formed on the lower leaf axils should be buried.

EXCESS LIGHT

Leaves appear bleached in the center and may have brown tips. To control this problem, shade should be given so as to reduce the light level to 1800-2500 foot-candles.

PLANT PROTECTION

Pests Aphids : Dimethoate (0.3 per cent)
Scale insects : Malathion (0.1 per cent)
Spider mites : Wettable sulphur (0.03 per cent)
Thrips : Malathion (0.1 per cent)
Diseases Anthracnose : Bavistin (0.1 per cent)
Leaf spot : Dithane m-45 (0.2 per cent)

Root rot : Captan (2 g/l) – soil drench
Bacterial wilt : Streptocyclin (200 ppm)

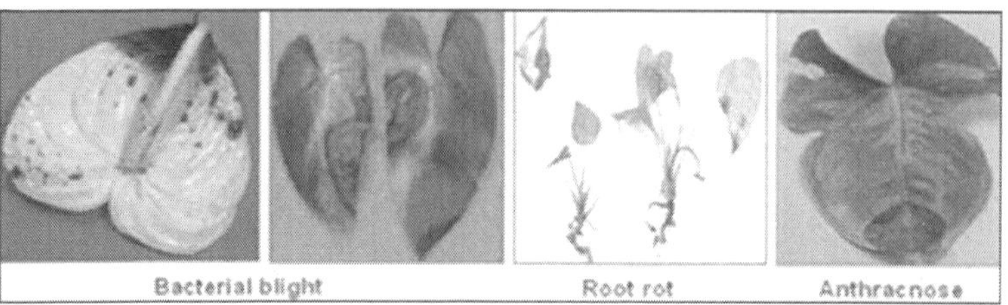

HARVEST

Harvest commences after 3 – 6 months of planting. Each leaf unfold will give out one flower. Flowers are harvested when the spathe completely unfurls and the spadix is well developed with one third of bisexual flowers got opened. Harvesting has to be done during cooler parts of the day *i.e.*) early morning or late evening. In general, the blooms are placed in water held in plastic buckets immediately after cutting from the plant. Delay in keeping in water allows air entry into the stem and causes blockage of the vascular vessels. Cut flowers after harvest should be shifted to pre cooling chambers in refrigerated vehicles having 2-4°C temperature as they deteriorate most rapidly at high temperature.

YIELD

An average 8 flowers/plant/year can be obtained.Post harvest technology:

1. Pulsing of flower stalks with BA 25 ppm for 24 hours improves shelf life up to 24.5 days as against 13.5 days in control
2. Packing the spathe with spadix in poly film (100 gauge) and covering the basal ends of the stalks with cotton dipped in BA improves shelf life up to 27.5 days
3. Holding solution: 8 HQC 200 ppm + sucrose 5 per cent increases vase life up to 30.5 days

DENDROBIUM ORCHID (DENDROBIUM SP.)

VARIETIES

Sonia 17, Sonia 28, Emma White, Sakura Pink. .

Fig. Sonia-17.

Fig. Sonia- 28.

Fig. Pravit White.

ORCHIDACEAE

Dendrobium species is a typical tropical orchid species suitable for Chennai and other coastal areas where the humidity is high.

Fig. Dendrobium Cultivation Under Shade Net.

CLIMATE

75 per cent green shade net with 70 - 80 per cent humidity, 18 - 28°C temperature and light intensity of 1500-2000 foot candles is ideal for growing this tropical orchid.

GROWING ENVIRONMENT:

75 per cent shade net house with 70- 80 per cent humidity, day temperature of 21 - 29oC and night temperature of 18 to 21oC is ideal for growing this tropical orchid. In high rainfall zones, the shade net house should be provided with a rainshelter.

PROPAGATION:

Division of clumps, keikis, back bulbs and tissue culture plants.

Containers and support : perforated earthen pots are ideal and the plants are staked with bamboo sticks.

Fig. T C Plants.

GROWING MEDIA:

Most common potting mixture consists of charcoal, broken pieces of bricks and tiles, coconut husk and fiber.

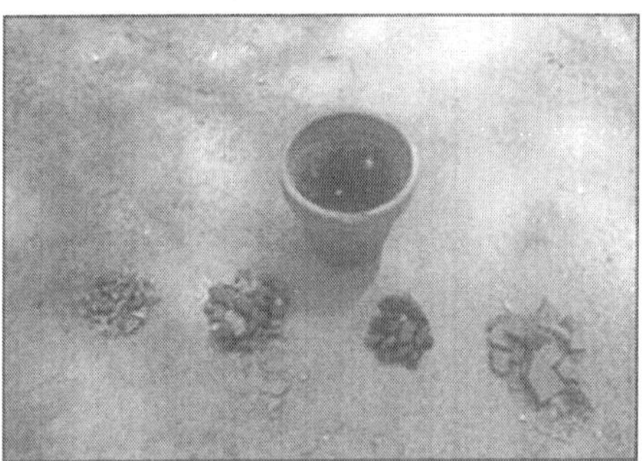

Fig. Growing Media.

Irrigation:

Mist or overhead sprinkler to provide water and to maintain humidity.

Growth Regulators

Foliar application of GA3 50 ppm at bimonthly intervals starting from 30 days after planting.

Repotting

Orchids need repotting regularly, usually every two to three years:

1. When the plant grows large and overgrows its container.
2. When the potting material deteriorates

The best time for repotting is when fresh roots emerge at the bases of the previous year's growth. In monopodial climbers, repotting or division has to be done when new leaf growth shows at the top and there is new root growth.

Nutrients:

Foliar application of NPK 20:10:10 @ 0.2 per cent at weekly intervals starting from 30 days after planting.

Fig. Foliar Feeding of NPK - 17:17:17 Complex

SPLITTING OR DIVISION OF PLANTS

Plant grown to a large clump with 2 or 3 old canes and new shoots, - divided before repotting. Each division - at least one old cane of two years' growth, one new shoot and some new roots. Pests: Snail and slug: Hand pick and destroy

Harvest

Fig. Spike Ready for Harvest.

Dendrobium flower fully matures only 3 or 4 days after it opens. Flowers are harvested when they are fully open as the flowers cut prior to their maturity will wilt before reaching the wholesaler. Immediately after harvest, the lower 0.75cm of the peduncle is cut off, and the flower is inserted into a fresh tube of water containing preservative. Harvesting the spike when 75 per cent of the flowers are open and remaining buds are unopen.

Post harvest handling:

Pulsing	:	8-HQC 500 ppm + Sucrose 5% for 12 hrs
Holding solution	:	$AgNO3$ 25 ppm + 8-HQC 400 ppm + Sucrose 5%
Wrapping material	:	50 gauge polythene with base of spikes dipped in 8-HQC 25 ppm

Yield:

8 - 10 spikes/plant/year

Pests:

Snail and Slug:

Hand pick and destroy them immediately.

Diseases:

1. Bacterial soft and Brown rot (Ervinia spp.) Foliar application with Streptomycin Sulphate @ 0.5 g + Copper Oxy Chloride @ 2 g/l.
2. Bacterial Brown spot (Acidovorax sp.) Foliar application with Streptomycin Sulphate @ 0.5 g + Copper Oxy Chloride @ 2 g/l.
3. Blackrot (Pythiumsp.and Phytothora sp.) Foliar application of Metalaxyl 2 g / lit. (or) Dimethomorph 50 per cent WP 0.5 g / lit.
4. Anthracnose – Foliar application of Thiophanate Methyl 2 g / l (or) Difenoconazole 0.5 ml/l

CUTFLOWERS-ECONOMICS OF PRODUCTION AND EXPORT

In an area of floriculture or flower production, Cutflowers have assumed prominent place in respect of (1) Selectivity or type of flowers, (2) Method of cultivation, (3) Marketing and (4)Final consumer use

SELECTIVITY OR TYPE OF FLOWERS

From amongst the wide range of flowers, only certain type of flowers are grown as cutflowers because of their special features, particularly long stem or stalk. For example, rose, carnation, gerbera, gladiolus, tuberose, anthurium, etc. There is also varietial preference for them according to the choice of consumers.

METHOD OF CULTIVATION

Open field cultivation has been a traditional practice, which is a relatively cheaper method. In modern "Hi-tech" method the cutflowers are grown in polyhouses/greenhouses requiring high capital investment. But the quality of flowers produced is superior, because inside climate or micro-climate such as temperature, humidity, light, ventilation etc is controlled. Even water application is also controlled. Even water application is also controlled. Therefore, the quality of flowers is better. They are uniform in size, colour, freshness etc. Moreover flowers can be produced throughout the year to meet the market demand-domestic as well as foreign. Since flowers are of better quality, they fetch higher prices.

ECONOMICS OF PRODUCTION

The polyhouses in which cutflowers are grown are of various sizes ranging from 500 sq.m.to 10,000 sq.m. (One hectare). They also differ in terms of cost as (a)low cost-Rs.125/m^2, (b) medium cost- ₹.500/ m^2 and (c) high cost-Rs.2000/ m^2, depending upon material used for construction and other facilities provided in them. An investment in a polyhouse of one hectare size with medium cost comes to almost ₹. 50 lakhs, which is quite high. Considering high initial investment, the Government of India has introduced scheme of subsidy from 10 per cent to 50 per cent.

Economics of cutflowers production (Roses) in a polyhouse of one hectare size is given below:

Sr. No.		Particulars	Amount Rs. in lakh
I		Capital Investment	
	1	Polyhouse structure	45.00
	2	Pre cooling and cold storage unit	15.00
	3	Refrigerated van	13.00
	4	Planting material	4.50
	5	Land value	2.00
		Total	79.50
II		Costs and Returns	
	1	Fixed costs	16.43
	2	Variable costs	29.82
		Total costs	46.25
	3	Total no. of flowers produced during 9 months (N0.)	6,76,170
	4	Per flower	
		Average cost (Rs)	6.85
		Average price received with no important (Rs)	12.28
		Average price received net ofimport duty (15%) (Rs)	10.44
		Net profit (Rs)	3.59

The items of fixed cost included interest on investment, depreciation on structure and transport and other equipment, and amortization of planting material. The items of variable cost included irrigation charges, fertigation, labour charges, managerial and supervision charges grading, packing, transport costs and air freight. This showed that inspite of high cost of cultivation of roses in polyhouses, their production is quite profitable due to export market.

QUALITY GRADES

Cutflowers are graded according to the length of stem or stalk, which varies from 5 cm to 120 cm. Longer the stalk better the quality and hence higher the price. Most commonly followed grading is designated as-

a. Short - stalk length – below 45 cm.
b. Medium- stalk length – 45 to 60 cm.
c. Long - stalk length – more than 60 cm.

Most of the cutflowers (50 per cent) were of medium stalk. Average price received per rose flower according to stalk length was ₹ .7 for short, ₹ . 16 for medium and ₹ . 20 for long stalk.

PERIOD OF HIGH DEMAND

In western countries, Valentine Day and Christmas festival are the periods of high demand and consequently of high prices. Per flower prices of roses were ₹ . 26 at Valentine Day, ₹ .17 at Christmas festival and ₹ .10 at other times. This trend in prices needs to be considered while planning cutflowers production.

Table. Another Study Gives Economics of Cultivation of Gerbera Flowers (Estimate for 2500 sq.m. Area that is 17500 Plants).

Sr. No.		Particulars	Amount
I	1	Fixed costs	3,50,000
	2	Variable costs	
		First year	7,63,150
		Second year	7,13,150
		Total	18,26,300
II		Returns	
	1	Total number of flowers per year @ 50 flowers per plant for 2 years (No.)	17,50,000
	2	Sale value @ Rs. 2.50 per flower	43,75,000
	3	Total cost of two years	18,26,300
	4	Net profit for two years	25,48,700
	5	Per flower	
		a) cost of cultivation	1.04
		b) sale price	2.50
		c) net profit	1.46

Per flower cost of cultivation of roses is much higher (Rs. 6.85) than that of Gerbera (Rs. 1.04). Production of Gerbera is also quite profitable.

MARKETING

Since cutflowers are of specific type and produced in polyhouses they are fresh and tender and since they are produced for specific purpose, great care is needed in their marketing *viz.* packing, handling, storage and transport. There should be minimum handling and transport should be quick with cooling and refrigeration facility. This is particularly necessary for cutflowers, which are produced for, export purpose.

FINAL CONSUMER USE

Final consumer use of cutflowers is different from other flowers. Their use is of more sophisticated nature in educated and well-to-do segment of consumers. Cutflowers are mainly used for preparing bouquets, which are used in functions and ceremonies to welcome guests, VIPs and to felicitate great utility and hence fetch high prices.

SUBSIDY SCHEME

Looking to very high initial investment in polyhouses and to encourage more floriculturists to undertake cutflowers production for export purpose, the Govt. of India introduced an incentive scheme of subsidy for construction of polyhouses. The details of the scheme are as follows.

Table. Thus the Subsidy Varies From 10% to 50% with a Ceiling to the Amount Upto ₹ . 1,00,000.

Sr. No.	Type of Greenhouse	Cost/ m^2	Max. area m^2	Total Cost Rs	Subsidy %	Amount of Subsidy available Rs.
1	Low cost	125	500	62,500	50	31,250
2	Medium cost	500	500	2,50,000	40	1,00,000
3	High cost	2000	500	10,00,000	10	1,00,000

3

Plant Maturity and Flowering Time

Many plants grow vegetatively for periods ranging from weeks to years and then flower autonomously, apparently without identiable environmental control. Flowering of 25–30 year-old bamboo is one such example: no environmental cue is known for this species. Perhaps it has its own built-in developmental clock which determines flowering time as in some annuals which flower autonomously. In contrast, other species may flower late due instead to inappropriate cultural or environmental treatments. In this instance, flowering may not occur irrespective of whether the juvenile phase has ended.

In some species, flowering occurs after the apex has produced a particular number of leaves. This apparent leaf counting may reflect an interplay between older leaves and the roots. In tobacco, for instance, proximity of the roots to the main shoot apex is critical. Plants remain vegetative until the shoot apex is more than ve to seven leaves above the roots or above a zone of experimentally induced root formation on the stem.

Extremely fast flowering without any apparent juvenility is seen in some desert annual plants. They may germinate and reproduce rapidly after rainfall, forming as few as two or three leaves and then flowering. The terminal shoot apex and all axillary apices may become floral. More often, however, such rapid flowering is restricted to either lateral or terminal meristem(s), leaving a second population of meristems avail-able for further growth and reproduction if favourable conditions persist .

With some agricultural crops bred for earliness of flowering, such as soybean and rice, early maturity may have resulted from a shortening of the juvenile phase (Evans 1993) rather than from changes in sensitivity to environmental cues. Thus, for some crop plants, duration of juvenility can influence chronological and developmental time from seed germination to flowering, regardless of other physiological controls of flowering.

As an adaptation for survival, juvenility is an advantage and a single gene controlling its duration is known inPisum . Embryonic flowering (Emf) may peform a similar role in Arabidopsis. As discussed later, several other floral-specic genes also influence aspects of this floral transition. In contrast to the abbreviated juvenile phase of annuals, perennials such as apple or mango have a juvenile phase often lasting ve to eight years. Various cultural and environmental manipulations including drought, nitrogen fertilisation, stem girdling, grafting and CO_2 enrichment can reduce this period in conifers. The juvenile period of some Eucalyptus species can also be shortened from two to three years to 9–12 months if grafted cuttings are exposed to cool inductive conditions and treated with an inhibitor of gibberellin biosynthesis. Endogenous gibberellin A_1 (GA_1) levels were lowered by this treatment so high gibberellin levels may be one component of prolonged juvenility in Eucalyptus.

FLOWERING TIME AND ENVIRONMENT: PHOTOTHERMAL INPUT

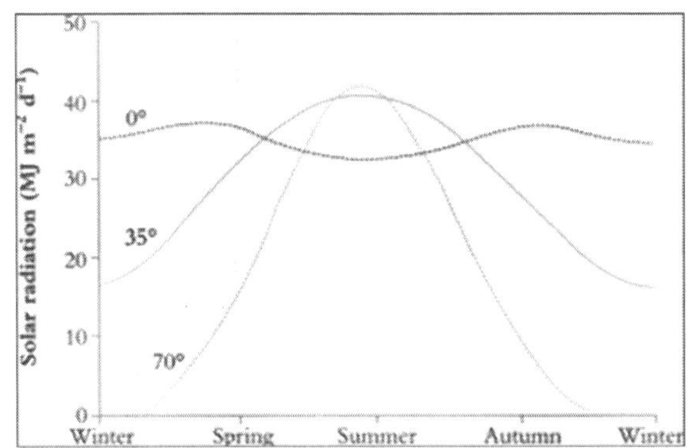

Fig. As the Seasons Change, Solar Radiation Incident on the Earth can Fluctuate Dramatically at Extreme Latitudes or Very Little at the Equator.

Environmental factors that limit plant growth may also pro-foundly influence flowering time. Suboptimal growth conditions may delay flowering and give an apparent ex-tended juvenile phase, and often light intensity, light duration and temperature are major limitations. Thus, a summation of both inputs (the photothermal sum) over all or part of the calendar year helps to characterise the growing season. Photo-thermal sums indicate whether there is adequate time from sowing to seed maturation for an annual crop or wild plant species. The yearly cycle of solar radiation highlights how this varies with latitude. There are losses due to cloud and to atmospheric interception. Of the remaining sun-light, the visible/photosynthetic component is about 45 per cent and the rest is 'heat'. The calculation of photothermal units integrates these heat and visible light inputs. For example, although daily photosynthetic flux at extreme latitudes may be high in summer, the growing season is extremely short.

Thermal sums (based on a heat sum above a 10°C base) have been used in the USA to predict the likely penalty in flowering time, and hence in yield, from growing long-season (late flowering) corn varieties at a higher latitude. To maintain yield, breeders have had to obtain lines with shorter growing seasons, in this case selecting varieties with more rapid early seedling growth and therefore re-quiring smaller thermal sums. Similar approaches with other crops such as soybean have used data from analysis of eld environments and controlled-environment studies.

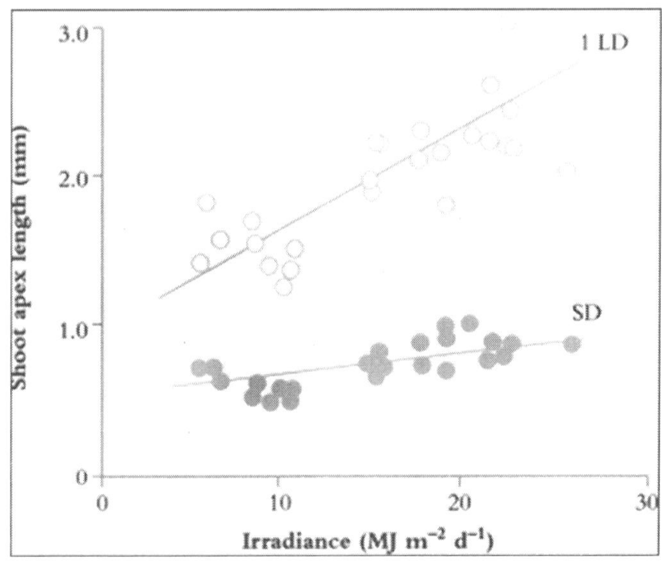

Fig. Effect of Seasonally Changing Total Radiation on Inflorescence Initiation in Lolium Temulentum Growing in a Fixed-Temperature Regime. Plants Either Flowered Ager a Single Long-Day (1 LD) Exposure Given at Different Times of Year of Remained Vegetative in short Days (SD). L. Temulentum is a Long-Day Plant with Apex Length e" 1 mm indicating Transition to a Floral State, Measured 21 d Following Floral Induction.

Photothermal responses for perennial crops are more complex, partly because flowering may relate to current and previous years' environmental conditions. Controlled-environment experiments help us unravel some of the interactions. In vines such as grape and kiwifruit, the extent of bud dormancy can be determined on cuttings taken from 'winter' canes and transferred to controlled-environment cabinets. This enables prediction of timing of eld budburst for each cultivar.

Another approach with perennial plants involves collection of eld flowering and temperature data over a number of years at different latitudes. For two ericaceous shrubs a heat sum model predicted flowering times at eight eld sites in Canada (Reader 1983) and similar heat sum relationships have been shown for another 15 species at 200 latitudinal sites in Alberta. The earliest spring flowering species had the smallest heat sum for flower opening. Information on climate and plant responses to the environment provides one way to estimate global re-productive potential. In equatorial zones, temperature and irradiance change less over the year and time of flowering may instead reflect seasonal rainfall patterns. In warmer temperate zones, early spring flowering and adaptation to intermediate heat sums can ensure reproduction prior to high summer temperatures and drought stress, but a second favourable climatic window is autumn. At high latitudes or at altitude, growth and flowering occur during midsummer.

Although these ideas can explain seasonality of flowering, photothermal relationships match best to the period of development up to flower opening (Reader 1983). They apply less well to floral induction, which is often a response to specic episodes of high or low temperature and/or to seasonal change in daylength. Assessment of such responses is best studied in controlled-environment chambers where each component can be varied independently. In this way we can reveal effects on flowering of seasonal changes in amount and duration of daylight, the 'photo' component of photothermal responses. The flowering response of the grass Lolium temulentum varies with irradiance at the time of exposure to a single inductive long day. Increase in photo-synthetic input is benecial but is not the major limiting factor for flowering. Rather, daylength (photoperiod duration) is the major determinant of flowering in this and many other species.

DAYLENGTH AND FLOWERING TIME

As long ago as 1914, scientists recognised that daylength regulated flowering time of hops (Humulus japonicus) and by 1920 two Americans, Garner and Allard, had demonstrated daylength control of flowering of many species. They termed the species either short- or long-day plants (SDPs or LDPs). SDPs flower in response to a decrease in daylength, that is, an increasing length of the daily dark period and a shortening photoperiod; LDPs flower in response to increasing photo-period. As well as causing flowering, daylength can also influence winter dormancy of buds, tuberisation, leaf growth, germination, anthocyanin pigmentation and sex expression.

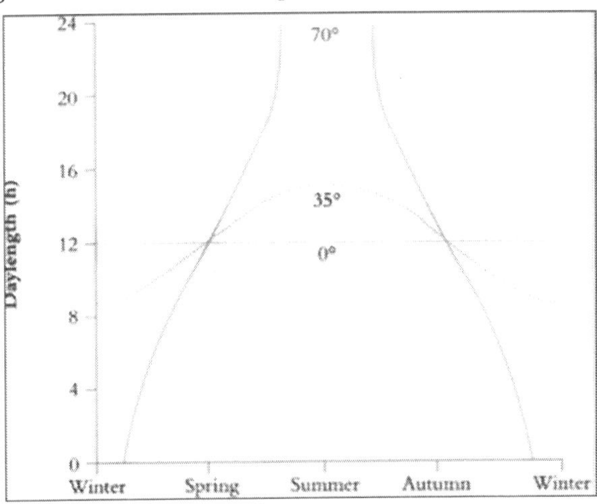

Fig. Seasonal Daylength at Various Latitudes. Values at Other Latitudes Fit Between those Shown.

Change in daylength is identical from year to year and so provides precise information on season. Thus a photoperiodic plant can time reproduction to avoid mid-summer drought, autumn cold or late spring frosts. Summer flowering at higher latitudes typically will involve a response to long days. In the tropics, daylength changes little, so selection pressure could be for daylength insensitivity or short-day response, provided plants could measure such small changes in daylength. Withrow (1959) calculated that to measure seasonal time to within one week required a 1–3 per cent precision in measurement of daylength. Only a 4–12 per cent precision was required for accuracy to within a month. In the tropics, a 1–3 per cent accuracy would mean distinguishing photo-periods differing by 7–21min around a 12h daylength. Remarkably, several species including some tropical plants do show such accuracy. In studies with rice, a tropical SDP, flowering occurred 30 to 50 d later when the photoperiod was increased by only 10min, from 11h 50min to 12h (Dore 1959).

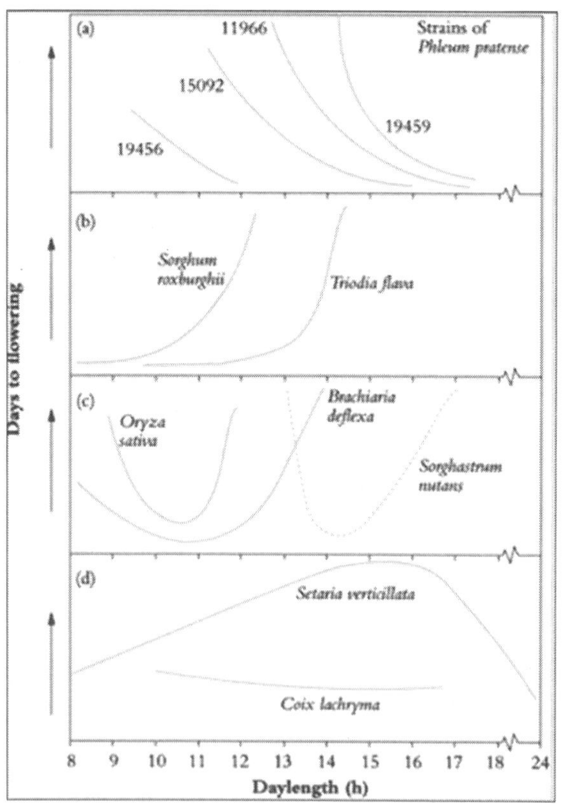

Fig. Control of Flowering by Daylength in (a) Several Strains of a Long-Day Grass, (b) Two Short-Day Grasses, (c) Three Intermediate-Day Grasses, (d) a Daylength-Indifferent and an Ambiphotoperiodic Grass.

Detection of daylength involves a photoreceptor called phytochrome. This pigment detects very low energies of visible light, especially red and far-red wavelengths. The con-sequence is that major daily and seasonal fluctuations in photosynthetic light intensity do not influence measurement of daylength. So sensitive is phytochrome that at latitudes up to 40° plants respond to twilight radiation for about 20min after sunset and before sunrise . At high latitudes, the midsummer sun may never set as far as phytochrome sensing is concerned. We return to discussions of phytochrome.

The duration of daily light/darkness which is effective for flowering may be very precise or very broad. Such contrasting patterns are along with typical long-day, short-day, intermediate, ambiphotoperiodic or day-neutral (indifferent) responses. Daylength-indifferent types represent less than 15 per cent of the 150 or so grass species reviewed by Evans (1964), although this proportion may be an underestimate as 'observed' day-neutral responses might not always be reported.

Within a species there can be large differences in photo-period response, as in the LDP Phleum pratense. The full range of daylength response types may even be found within a single species. For example, in a controlled environment study of 30 ecological races of the Australian grass Themeda australis, Evans and Knox (1969) found that low-latitude strains, from 6° to 15°S, behaved as SDPs. Races from more southerly origins to 43° were LDPs with some responsive to vernalisation. This ecotypic variability exemplies heritability and adaptability of environ-mentally responsive flowering and appears to have aided reproductive success of Themeda. If the species migrated to Australia via Asia and New Guinea, it would probaly have adapted from a short-day response to day neutrality or sensitivity to long day and to vernalisation.

Some plants will flower after just one cycle of the appropriate daylength. Cocklebur (Xanthium strumarium) and Japanese morning glory (Pharbitis nil) are classic examples of SDPs responding to one short day, or more correctly, one long night. Similar single-cycle responses are found for LDPs such as Lolium temulentum. Other species require several days (*e.g.* soybean, strawberry) or weeks of exposure to the appropriate daylength (*e.g.* Geraldton wax, chrysanthemum).

In some plants a sequence of short days must precede long days (SLDP) as for some clovers (*e.g.* Trifolium repens) and grasses (*e.g.* Poa pratensis). Conversely, some species respond as long–short-day plants (LSDP) in-cluding Aloe, Bryophyllum and some mosses and liverworts. Some dual photoperiodic responses may be satised simultaneously so that flowering is best at intermediate daylengths (*e.g.* some sugar cane genotypes). The converse is also known, ambi-photoperiodic response, with best flowering at either short or long days but not at intermediate daylengths. Separation in time occurs in some grasses which respond to short days for primary induction leading to a microscopically visible inflorescence but later to long days for subsequent development to anthesis (Heide 1994).

LOW TEMPERATURE AND FLOWERING TIME

VERNALISATION RESPONSES

Although growth is limited by low temperature, scientists in the mid-nineteenth century recognised that floral initiation of many species requires exposure to cold. For a temperate cereal such as wheat, low-temperature exposure of imbibed grain caused winter lines to flower like their spring wheat counter-parts. We term this response vernalisation, meaning 'to become spring-like'.

Vernalisation-responsive species include winter annuals, biennials and perennials. Many are also LDPs including some grasses and species with a rosette growth habit. Effective temperatures for vernalisation range between -6°C and 14°C, with most temperate species responding best between 0°C and 7°C. In all cases, these temperatures are below those optimal for growth. Floral primordia are sometimes initiated during the cold period, as in brussels sprout, turnip, stock and bulbous iris. Alternatively, cold treatment is a preparatory phase enabling later initiation of flowers.

Generally, prolonged exposures of one to three months are required for vernalisation but this varies with temperature and species. However, as with photoperiodic species, some respond to a single cold day, for example chervil. In Geum, the vernalisation period depends on meristem location, ranging from two to three months in axillary meristems to one year for the terminal apex. Heterogeneity of floral response of meristems has clear adaptive benets, whether for perennation as with Geum or for opportunistic responses to rainfall as for desert ephemerals.

As with photoperiodism, dependence of flowering on vernalisation changes with latitude. For example, a vernalis-ation response appears only in high-latitude ecotypes of Themeda australis and is likewise more important for species and ecotypes from higher altitudes. European thistle (Cirsium vulgare) collected from the Mediterranean to Scandinavia exhibit vernalisation requirements predominantly in lines from colder, more northerly sites . In addition to latitude effects in the grass Phalaris aquatica, there is a superimposed altitudinal cline.

Leaves sense photoperiod, but perception of low tempera-tures resulting in vernalisation responses can be by the shoot apex instead. Chilling of leaves is usually ineffective (Bernieret al. 1981). However, cold-treated leaf cuttings of species such as Lunaria and Thlaspi arvense, and even chicory root explants, regenerate plants which flower without further vernalisation . One hypothesis is that vernalisation responses may be initiated only at sites with potential for cell division, that is, meristems or regenerating tissues. On the other hand, in pea and sweet pea, there is clear evidence of transmission of vernalisation signals across graft unions. In these experiments, perception of cold must have occurred in cells other than those in the responding shoot apex. These species also exhibit normal shoot apex vernalisation responses, so there can be two different mechanisms of low-temperature sensing.

The presence of water and metabolic activity are essential requirements for vernalisation. We deduce this from vernalis-able species which can respond during seed germination. Radish seed, for example, cannot be vernalised when dry or in a nitrogen atmosphere. The vernalised state is quite stable in seeds of some species: they can be dried after cold treatment, even stored for long periods, and then sown without loss of response. However, particularly with marginal vernalisation, temperatures immediately following often need to remain below 25°C to prevent devernalisation. High temperature up to 40°C for a few days sometimes annuls a preceding cold exposure . Indeed, devernalisation every summer may reset the flowering of perennial plants so that they require renewed vernalisation each winter.

Photoperiod requirements post-vernalisation are diverse. Many winter annuals or biennials require long days following vernalisation. For example, vernalised Hyoscyamus will not flower under short days but under long days promptly forms flowers, even with 300 short days between vernalisation and induction. In contrast, sensitivity of spinach to inductive long days is altered following cold treatments with a shortening of the critical day length from 14h to 8h. A few cold-responsive plants, such as chrysanthemum, require short days after vernalisation.

The genetics of vernalisation range from simple to very complex depending on the species. For example, a single locus distinguishes the biennial, cold-requiring strain of Hyoscyamus from its annual counterpart. By comparison, vernalisation of hexaploid wheat involves at least three loci (Vrn 1, 3 and 4), probably reflecting its genetic complexity.

Pea and Arabidopsis normally respond both to photoperiod and to vernalisation. Of the many late-flowering mutants known, some are vernalisation responsive, including gigas (gi) in pea and luminidependens (ld) inArabidopsis. There are also vernalisation-unresponsive and early-flowering mutants. One simple explanation is that the wild-type products of some of these genes are inhibitors of floral induction or initiation or, conversely, stabilise vegetative growth.

Vernalisation may involve decreased DNA methylation allowing activation of suites of genes including some involved in synthesis of gibberellins. For example, extending the earlier work of Hirono and Redei (1966), Burnet al. (1993) found that vernalisation-responsive late-flowering mutants of Arabidopsis treated with the demethylating agent 5-azacytidine flower earlier than unvernalised controls. From this result, they concluded that demethylation occurs during vernalisation and leads to selective derepression of genes required for flowering.

COOL TEMPERATURE RESPONSE

In addition to classic vernalisation responses, there are many reports of species, especially from warm climates where near-freezing temperatures are infrequent, which flower if exposed to temperatures from 10°C to 20°C. For some tropical fruit crops (*e.g.* mango, avocado, lychee, longan), especially those grown in the subtropics (latitude 23°–30°) where substantial seasonal temperature changes occur, floral induction results from exposure to night temperatures of 10–15°C. Because tropical species are relatively under-researched compared with their temperate counterparts, physiologists have yet to decide whether these cool responses have similar mechanisms to temperate vernalisation but are adapted to a different temperature range. Another possibility is that flower initiation and development are blocked/ reversed by higher tempera-tures, so low temperature could merely be a passive condition permitting expression of an innate capacity to flower. This may be the case for Acacia and rice flower but for Pimelea ferruginea, which flowers if exposed to temperatures below a daily average of 16–18°C for ve to seven weeks, the response is inductive and higher temperature does not cause loss of developing flowers .

WATER STRESS AND NUTRITION

In some species including Lolium, Pharbitis and Xanthium, floral induction and development are blocked by water stress. For Lolium, an 8h stress inhibited flowering only if given at the time of the long day, not one day before or after. Shoot apex abscisic acid content increased transiently up to 10-fold in association with the brief water stress (King and Evans 1977). Furthermore, ABA inhibited flowering if applied at the time of the long day. Later in flower development, water stress or ABA application can result in sterility in wheat. The problem is morphologically aberrant pollen, but seeds are still set if plants are hand pollinated. By contrast, positive responses of flowering to water stress are also known. For the geophyte Geophila renaris, growth under water-limited conditions for two months causes flowering. Similarly, water stress coupled with enhanced photosynthetic conditions, high tem-perature and gibberellin application can cause precocious flowering in some conifers (Pharis and King 1985). In mango trees grown in the tropics with little temperature variation, seasonal flowering appears to be promoted by water stress during the dry season. This may relate to trees having an extended period of suspended growth during which ability to flower gradually develops, for example as a result of accumulation of stored carbohydrate.

Nutritional status of plants has little direct influence on floral initiation, although in many species there are effects on flower number and on fruit and seed development. For example, pollen fertility in wheat is reduced by excesses and deciencies of trace elements including copper and boron (reviewed by

Graham and Nambiar 1981). In strawberry, plant size and fruit and flower number increase as nitrogen supply is increased (Guttridge 1969), but the supply of nitrogen during early stages of flower initiation may enhance vegetative growth not flowering. Such complex responses make it dif-cult to argue that transition to flowering requires low-nitrogen status coupled with enhanced carbon supply.

Numerous studies have failed to demonstrate an inverse relationship between nitrogen supply and flowering and, as noted above, there are often positive effects on floral development. Perhaps a unique response to nitrogen is the dramatic increase in flowering of apple supplied with nitrogen but only if supplied as ammonia (Grasmanis and Leeper 1967). Overall, mineral nutrients, while essential for growth, may not specically regulate flowering.

ENVIRONMENTAL AND SEASONAL SYNCHRONISATION OF FLOWERING

SPECIES IN THEIR NATURAL ENVIRONMENTS

Control of seasonal flowering time may be as simple as the acquisition of a long-day or short-day photoperiodic response, or of both as in LSDP where exposure rst to long summer days is essential to guarantee flowering in the short days of autumn. Alternatively, floral development may occur in spring when both temperature and irradiance increase rapidly to permissive levels. A vernalisation requirement allows for spring flowering, or for summer flowering when combined with a long-day response.

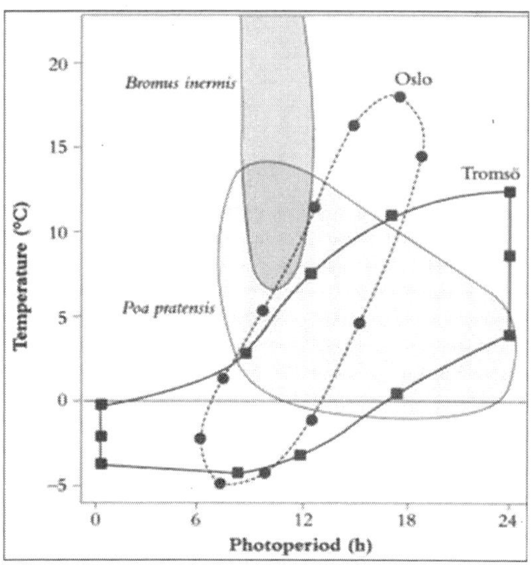

Fig. Climate Phototherms for Tromso (69°39′N) and Oslo, Norway (59°55′N). Mean Monthly Temperature Versus Photoperiod Together with Optimum Areas for Primary Induction of Flowering of Bromus Inermis and a High-Latitude Species Poa Pratensis.

Often, a combination of short day then long day, as well as temperature, is important in synchronisation of flowering of perennial grasses (Heide 1994). Comparison of environmental tolerances of Bromus inermis, a species adapted to lower latitudes, and Poa pratensis, an arctic–alpine species, highlights how these inputs determine survival. For flowering, both species require short-day or low-temperature exposure followed by long days. The short-day response is strict in Bromusand, because of intolerance to low temperatures, it will never flower at the high latitude of Tromsö (69°39¢N), as shown by its climate phototherm.

The response of Poa, by contrast, overlaps an arctic phototherm (Tromsö) but this species is intolerant of the higher summer tem-peratures at lower latitudes. Dual induction responses also enable high-latitude-adapted species to initiate inflorescence primordia in autumn short days. The outcome is to maximise the number of summer days available for seed development because anthesis proceeds rapidly in the following summer long days, even in the short, cool arctic growing season.

Field to nursery transplantations have often demonstrated environmental influences on flowering, as noted above for vernalisation of Cirsium arvense. Alternatively, controlled en-vironment studies of the type used by Evans and Knox have revealed ecotypic adaptation of flowering in Themeda. Rarely have the two approaches been combined. Either photothermal models have been used to assess eld flowering data or laboratory environmental response proles have been incorporated into empirical models predicting eld response. However, with Pimelea ferruginea grown simultaneously in controlled environments and in the eld over winter , there was a close match between effective temperatures for flowering in the eld and laboratory. In addition, evidence for adaptation to small (4°C) temperature dif-ferences came from a high-latitude ecotype from 31°S which was unable to flower when transplanted to the warmer extreme of the species distribution (28°S).

PREDICTING FLOWERING TIME OF ELD CROPS

Phototherms only broadly dene the tolerance of a species to its environment. A more denitive approach uses rates of response of flowering to photoperiod and temperature based on constants derived from controlled environments. Threshold limits are also imposed to constrain models to response envelopes of the sort. In a broader study , six crop species (soybean, cowpea, mungbean, chickpea, barley and lentil) sown at different latitudes and times flowered in the eld at times which correlate well with those predicted from a simple linear additive model . However, such models make no allowance for effects of light intensity and extreme conditions outside the threshold limits which can be important for flowering, for example vernalisation or warm temperatures.

COMMERCIAL NURSERY FLORICULTURE

Prior information on environmental response has been crucial to nursery production of potted flowering plants including the SDPs chrysanthemum and poinsettia. However, there may be inevitable compromises in some of the complex protocols required for commercial production of an Australian SDP, Geraldton wax. Its critical photoperiod is about 13h, so the maximum tolerable daylength would be about 12h from sunrise to sunset plus 20min each pre-dawn and twilight (Dawson and King 1993). Thus, in summer, glasshouse black-out curtains are used to maintain the inductive short day, but this is obviously not an option for eld-grown plants. Glass-house summer temperatures exceeding 35–40°C, well above the optimum for the species, are another problem. As a comparison, optimal mean daily temperature for chrysanthemum is about 21°C . Con-sequently, greenhouses are often shaded to avoid costly cooling, but then lower photosynthetic input may result in poorer flowering.

FLOWERING OF WOODY HORTICULTURAL SPECIES

Prolonged juvenility of woody species is a problem for growers and breeders of tree and vine crops. However, there are so many uncontrolled variables in the eld that it can be difcult to identify the inductive factors. Yields can be severely depressed by inappropriate timing of practices such as pruning, irrigation and fertilisation. Furthermore, inductive conditions may be required for several months. One solution for mango, lychee, olive and citrus has involved the use of controlled environments and 'mini' plants grown from cuttings. These showed that cool temperatures were required for induction, a response similar to Pimelea and many other ornamental and woody species.

For some species, microscopic examination of shoot meristems has augmented our ability to make decisions on practical management of flowering. For example, in kiwifruit (Actinidia) and stone fruits (Prunus spp.) floral induction occurs in the previous growing season, whereas in many subtropical species no initiation takes place until winter. In the case of kiwifruit, it was discovered that late summer pruning was removing many of the floral apices .

Clearly, knowledge of environmental effects on flowering has been essential for development of nursery, orchard and agricultural crops. Particularly for eld crops, breeders have selected for day-neutral responses. For glasshouse crops, genotype and environment have often been altered. The future offers many opportunities for applying our knowledge of daylength and photothermal responses.

THE PROCESSES OF FLORAL INDUCTION AND INITIATION

Following the discovery of photoperiod-regulated flowering, there soon followed evidence of leaves as photoperiod sensors, of a timekeeper involving endogenous circadian rhythms, of transmissible florigenic signals and of a resulting cascade of developmental changes at the apex. Although sometimes used loosely, it has long been clear that the term 'flowering' embraces an amazing series of signalling systems and developmental transitions. Photoperiodic induction refers to photoreceptor-driven, leaf-specic processes. Flower initiation at the apex is now divided into floral evocation and floral differentiation; evocation describes the early processes occurring at the apex before irreversible commitment and differentiation of flower primordia. Although the term 'florigen' was coined initially, there may be multiple transmitted florigenic stimuli so 'floral stimuli' or 'florigens' are more appropriate.

PHOTOPERIOD AND LEAF PHOTORESPONSE

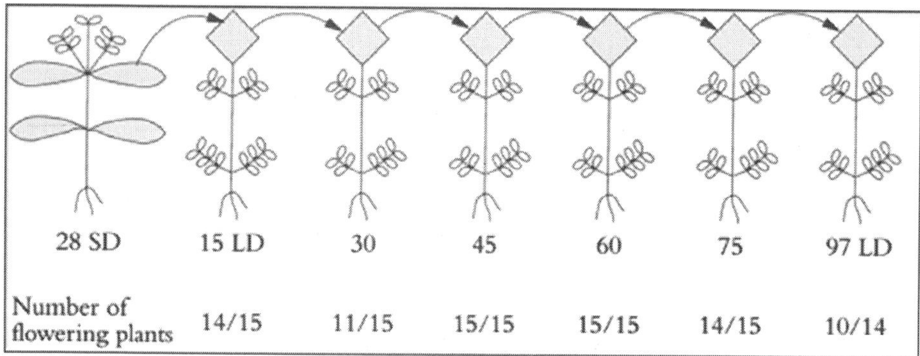

Fig. A Permanently Inductive State can be Demonstrated for Leaves of Some Photoperiodic Species. After 28 d of Short Days (SD) a Leaf of Perilla Returned to Long Days (LD) will Continue to Produce Graft-Transmissible Flowering Stimulus for at Least 97 d, Involving its Successive Grafts of the Same Leaf to Vegetative, Long-Day-Grown Receptor Plants.

Sensing of photoperiod requires photoreceptor pigments and a responsive organ. Elegant experiments involving selective light exposure of different parts of the plant conrmed that the leaf blade is the photoresponsive site. Defoliated plants show little or no photoperiodic response and direct illumination of the shoot tip is mostly ineffective. A leaf, once photoperiodically treated, may be permanently changed. Leaves of the SDP Perilla, for example, exhibit a remarkable permanently induced state to the extent that a single leaf is capable of causing flowering when grafted in sequence to six vegetative receptor plants over a period of 14 weeks.

There are at least three plant pigments that could regulate photoperiodic flowering responses: chlorophyll via photosynthesis, phytochrome and the blue light receptor. Photosynthetic input will enhance flowering as shown earlier for the LDP Lolium . Measurements of shoot apex sugars show that increased photosynthetic sucrose supply to the shoot apex may be important, but on its own it is insufcient. The primary requirement is instead for activation of phytochrome. For example, Lolium can flower in response to a single long day extended with non-photosynthetic light. Far-red-rich wavelengths from tungsten lamps are more effective than red-rich wavelengths from fluorescent lamps, and this is typical for LDPs. For another LDP, Arabidopsis, involvement of phytochrome in flowering is revealed by a brief (10 min) end-of-day exposure to pure far-red (FR) light which promotes flowering with classic R/FR photoreversibility. What in perhaps surprising, considering the range of phytochrome mutants inArabidopsis, is that none of the mutants presently known for phytochrome A or B delays flowering.

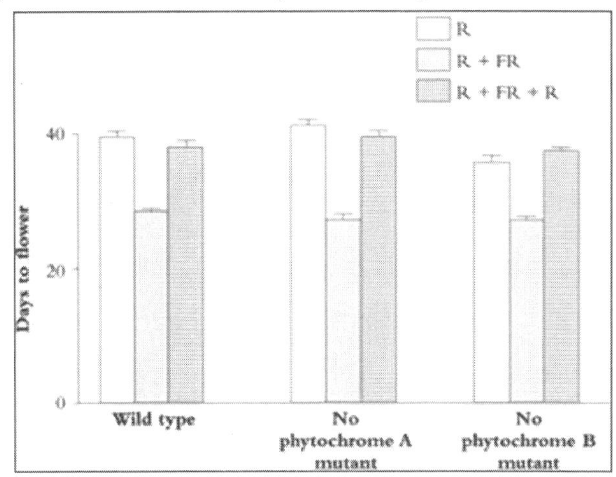

Fig. Photoreversible, R/FR Regulation of Flowering in the LDP Arabidopsis by Either Light-Stable Phytochrome B or Light-Labile Phytochrome A.

Photoperiodic treatment

Flowering response

SDP LDP

Flowering Vegetative

Vegetative Flowering

Vegetative Flowering

24 h

Light Darkness

Fig. Effect of Photoperiod and Night-Break Interruption on Flowering of SDPs and LDPs. The Night Interruption may be Less than 5 Min of Very Dim Light, as in Some SDPs, or May Require Prolonged (1-2 h) Exposures, as in some LDPs.

Phytochrome's role in flowering in SDPs relates to increases in the duration of the dark period. Light in the middle of the long inductive dark period (a 'night break') inhibits flowering of SDPs — they experience a 'pseudo' long day.

Conversely, night breaks may promote flowering of LDPs. For SDPs, the night-break duration may be amazingly brief (1–300s) and the response often shows R/FR photoreversibility. Other evidence from action spectra emphasises the importance of red wavelengths of light for SDPs in contrast to the response to far-red for LDPs.

PHOTOPERIODIC TIMEKEEPING

Accurate measurement of daylength for control of flowering requires a 'photo' response via a photoreceptor, and a measure of 'period' generally involving a circadian, rhythmic, timer. Circadian, meaning 'about a day', refers to the natural period of these rhythms often being not exactly 24h. In the absence of external stimuli, most rhythms manifest as free-running circadian cycles. However, the timing of dawn and/or dusk entrain the rhythm to synchronise with exact 24h cycles and hence provide an accurate daily clock used by both SDPs and LDPs. The currently favoured explanation of photoperiodic timekeeping involves rhythmic biochemical processes.

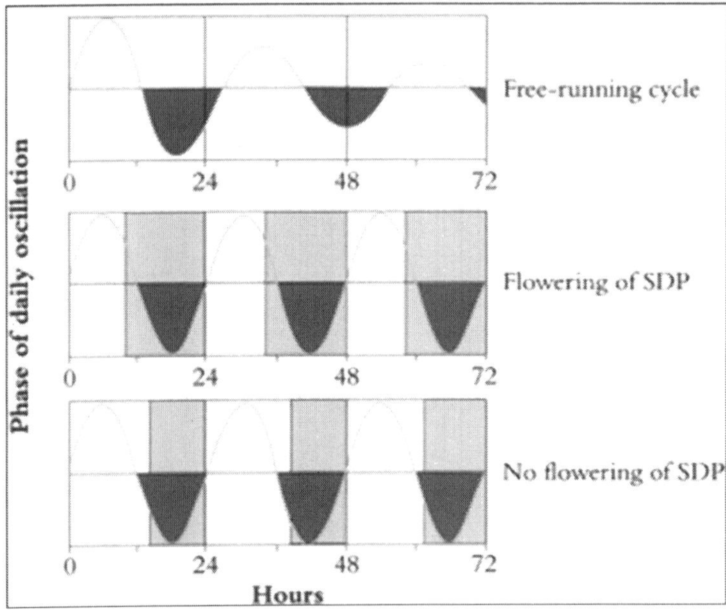

Fig. Daily Light/Dark Cycles (Empty/Light Grey Areas) Phase and Entrain a Free-Running Circadian from (Circa-Diem, Meaning 'About a Day') Oscillation to an Exact 24 h Cycle. It is Proposed that One-Half of the Cycle Tolerates Light with the Other Half (Dark Grey Portion) being Intolerent. Thus, for the SDP, Flowering is Only Permitted with Long Dark Periods. However, the Duration of Light and Darkness are Both Crucial Components of Time Measurement.

In addition, phytochrome is clearly involved, but may not act as an instantaneous on/off switch with respect to the light/dark cycle. Phytochrome is rapidly activated in light but on return to darkness there can be a slow (~ 0.5 to 4h) delay in disappearance of active phytochrome (the Pfr form) as it is degraded or decays back to the inactive Pr form. The consequence may be an offset between when it is actually dark and when the plant perceives it is dark. In the 1950s, Borthwick and Hendricks proposed that this natural offset, acting like an hourglass, accounted for photoperiodic time measurement in flowering . Nowadays, the hourglass theory is often dismissed, especially as it would be limited to measuring dark periods only up to 4h. How-ever, it does provide a rational explanation of flowering of SDPs exposed to an extended long dark period and may well be a necessary component of photoperiodic timekeeping but perhaps not the limiting factor. There may also be an essential stabilisation period after Pfr decay during which other forms of timing may occur.

Although daily light/dark cycles set the phase and entrain 24h rhythms, this does not explain photoperiodic control of flowering. For example, there are distinct phase settings of leaf movement rhythms for the SDP Pharbitis nilwhen in long or short days, but flowering is stimulated only by short days. In 1936, Bünning deduced that there is a second, additional, light response allowing or preventing expression of the rhythm. The phase of the rhythm imposes or determines sensitivity of flowering to this second light input. The consequence is that, depending on daylength, light may or may not be synchronised with the dark-requiring part of the rhythm and so flowering is either prevented or allowed.

Other rhythms have been revealed at the genetic and molecular levels. For example, Arabidopsisplants transformed with a luciferase gene for bioluminescence coupled to the promoter sequence for a clock-regulated plant gene gave a simple, visually assayed, indicator rhythm which was then used to screen for period length mutants . None of the mutants influenced flowering response, so it appears that there may be several independent clocks operating.

FLORAL STIMULI AND INHIBITORS

The diverse environmental influences on flowering make it unlikely that plants possess a simple, unique regulatory signalling system. At least for photoperiod responses, grafting experiments indicate the presence both of transmissible promoters and inhibitors. However, isolation of florigenic chemicals from induced plants remains at a preliminary stage. We are still uncertain whether the floral stimulus (or inhibitor) is a single compound, a complex of compounds, whether it is photoperiod class specic, species specic or more universal. Grafting experiments have conrmed that leaves produce photoperiodic stimuli that are transmitted to the shoot apex, as discussed earlier for Perilla. For several long-day and short-day species, pre-induced, grafted leaves or leafy shoots cause flowering of vegetative recipient plants held in non-inductive conditions. Intriguingly, grafted leaves from day-neutral species may even be effective donors to LDPs or SDPs held in non-inductive photoperiods. In a few cases, such as Sedum spectabile (LDP) and Kalanchoe blossfeldiana (SDP), interspecies grafts have also been successful. This tells us that, despite photo-periodic differences, there may be common stimuli or common perception by the apex of different stimuli.

Many unsuccessful, frustrating attempts to extract and identify flowering stimuli have led florigens sometimes to be called hypothetical, non-existent or the holy grail of plant physiology. In addition to the tobacco extract, some positive results have also been reported for the SDP Pharbitis nil . In both studies, there was activity only in extracts from induced plants. Importantly, there was no activity for extracts of non-induced long-day leaves or their phloem exudates. We predict from experiments measuring speed of transmission that the signal moves in the phloem but no florigen has been chemically identied. The identity of inhibitory compounds is a further mystery. The main evidence for floral inhibitors comes again from grafting studies, for example in day-neutral tobacco. When grafted with an LDP tobacco, Nicotiana sylvestris, the day-neutral line flowers late if the graft partner is in non-inductive conditions; we deduce that it is producing an inhibitor that can pass across the graft union. The converse experiment with the long-day partner in inductive days led to early flowering of the day-neutral plant, so there is also a transmitted promoter. However, Maryland Mammoth, a short-day tobacco, lacks the graft-transmissible inhibitor, indicating how difcult it is to unravel the complexities of signalling.

HORMONAL INVOLVEMENT

One reason for considering a role for plant hormones in the regulation of flowering is the frequent reports that their application dramatically alters flowering. However, cor-elations with altered endogenous hormone levels are not always evident, for example in the case of ABA content during floral induction inLolium. By contrast, gibberellin application can cause flowering particularly of rosette plants. It may replace a need for vernalisation or long days in control of bolting and flowering (Lang 1965) and, endogenous gibberellin content may also increase following environmental changes that lead to flowering. Some commercial uses of hormones have followed. For example, ethylene synchronises flowering and fruiting of bromeliads and is used worldwide for pineapple production. Conversely, inhibition of flowering of sugar cane by ethylene is practised in Hawaii where yield is greater if flowers do not develop (Moore and Osgood 1986).

With some ornamental species such as Spathiphyllum, most commercial growers use gibberellin because one application halves the time to flowering from six to three months. This early flowering is probably not related to juvenility, which is sometimes extended by applied gibberellin as in ivy (Hedera sp.) and shortened inEucalyptus nitens when gibberellin levels are lowered. After treatment with paclobutrazol, which blocks gibberellin biosynthesis, grafted seedlings flower massively and three to ve years earlier than normal. Yet we nd there are no generalisations. For conifers, high gibberellin level may overcome juvenility and applied gibberellins, in combination with harsh cultural conditions, allow flowering at one to two years rather than after 10 to 20 years.

For some non-rosette species, long days and/or vernalisation can lead to rapid increases in gibberellin content and inhibition of gibberellin biosynthesis may also block or delay flowering, which further suggests a link between gibberellins and normal reproductive responses. In species with no juvenile phase, gibberellins may replace the need for long days or vernalisation. For example, in the LDP Arabidopsis, a dwarf mutant (ga1-3) which is blocked in gib-berellin biosynthesis, flowers later than its wild type. In short days, some of these mutant plants may never flower unless treated with gibberellin. On the other hand, vernalisation fails to stimulate flowering. Evidence against a role for gibberellins comes from the normal flowering of dwarf genotypes of many species (*e.g.* pea, corn, wheat, rice) which are blocked in gibberellin biosynthesis or in capacity to respond to gibberellin (Reid and Howell 1995).

Gibberellins can instead be inhibitory, especially for some perennials, including Fuchsia,Bougainvillea, mango and citrus, and also for species such as strawberry. Other gibberellins are known which can stimulate flowering without affecting growth. A more extreme response is seen from some novel synthetic gibberellins which can even act as growth retardants while still retaining ability to promote flowering .

Complex relationships also exist between cytokinins and flowering. In the LDP Sinapis, endogenous cytokinin levels increase up to three-fold in long days. Applied cytokinin, however, induces only a partial flowering response . There can also be indirect effects as found in Pharbitis nilwhere cytokinins can alter assimilate distribution to give either inhibition or promotion of flowering .

We know much less about genetic and molecular events around the time of floral induction. Beginning with a late flowering mutant in Arabidopsis, a gene, Constans, has been identied whose expression is upregulated by long days and which may be one step in the sequence to florigens. Manipulation of phytochrome genes influencing flowering has also provided information on photoperiodic processes in leaves. In the future, we can expect to nd links to timekeeping genes which influence endo-genous rhythms. Analogous genes have been isolated from other organisms including Neurospora and Drosophila.

CLIMATE CHANGE AND THE FLOWERING TIME OF ANNUAL CROPS

A CHANGING CLIMATE

The recent fourth assessment report of the Intergovernmental Panel on Climate Change (IPCC, 2007a, b) provided clear evidence of changes in climate due to human activities. The concentration of greenhouse gases in the atmosphere has progressively increased over the last century or so. For example, [CO_2] has increased from pre-industrial levels of 280ppm to 379ppm and mean temperature has increased by 0.76°C over the same time period. Recent temperature changes have been particularly marked, such that the warming trend in the last 50 years has been 0.13°C per decade, nearly double that of the preceding 100 years. Projections to the end of this century suggest that mean global temperature will increase by 1.8–4.0°C (range 1.1–6.4°C), depending on the greenhouse gas emission scenario, accompanied by changes in rainfall patterns and an increase in climate variability . Such climate changes are expected to have far-reaching impacts on ecosystems worldwide.

Crop production is inherently sensitive to variability in climate. Some of the early studies of the impacts of climate change on crops highlighted the importance of changes in crop development at warmer temperatures in determining the impact of climate change on crop yield. For example, the yield of wheat declined by <"5–8 per cent (Wheeler et al., 1996) or 10 per cent per 1°C rise in mean seasonal temperature. The timing of anthesis and grain maturity was earlier at warmer temperatures in both studies, thus shortening the duration of growth and reducing grain yield. Under climate change, other factors, such as the enhanced rate of net photosynthesis at elevated [CO_2], will offset to some extent such decreases in yield due to temperature warming, and changes in precipitation patterns and the frequency of extreme weather events will further complicate impacts on crop yields. Nevertheless, it is clear that the impacts of climate change on crop productivity will be greatly influenced by how climate affects the rate of crop development, and hence the timing of crop growth.

The timing of flowering, a critical stage of development in the life cycle of most plants when seed number is determined, is important for adaptation both to the abiotic stresses of temperature and water deficit, and to biotic (pest and disease) constraints within the growing season. For example, in many annual crops, brief episodes of hot temperatures (>32–36°C) can greatly reduce seed set, and hence crop yield, if they coincide with a brief critical period of only 1–3d around the time of flowering (Matsui et al., 1997; Vara Prasad et al., 2000; Wheeler et al., 2000; Jagadish et al., 2008).

Therefore, the moderation of crop development will be critical to the impacts of climate change on yield in two ways: through determining the season length, and hence the availability of radiation, water, and nutrient resources for growth; and by affecting the exposure of the crop to climate extremes. Adaptation to moderate changes in climate that influence temperature, season length, and planting dates, as well as the occurrence of abiotic stress, can be achieved by selecting varieties with appropriate flowering times and crop durations. Farmers (landraces) and plant breeders (cultivars) have very successfully selected/manipulated life cycle duration and phenology to maximize the range of environments in which crops grow as well as their yield at least for current climates. A major challenge for crop improvement is how to plan for future climate change.

The genetic and environmental moderation of the timing of flowering is therefore central to the responses described above. The timing of flowering within a season is largely determined by responses to temperature and photoperiod, and in whole plants at suboptimal temperatures these quantitative responses are reasonably well understood. However, responses to temperature and photoperiod at supraoptimal temperatures are poorly understood—though clearly these will become more important as the frequency of high temperature events increases under projected climate change. In recent years molecular biology has also greatly contributed to our understanding of flowering gene pathways, although the effects of temperature and temperature×photoperiod interactions on these pathways have not been studied.

In this review studies of phenology in annual cropping systems, mostly with cereals and legumes, over the last 50 years that looked for evidence for changes in phenology in the recent past are first considered. Secondly, whole-plant responses of flowering to CO_2, temperature, and photoperiod are described, with particular emphasis on responses and interactions at high and supraoptimal temperatures. Thirdly, how genotypic variation in responses to temperature and photoperiod may be exploited to provide adaptation to climate change, and how this is simulated in climate change impact studies, is examined. The review concludes by identifying some key knowledge gaps in current understanding of the impacts of climate change on the flowering time of crops.

PAST CHANGES IN PHENOLOGY OF CROPPING SYSTEMS

Earlier flowering and maturity have been observed and documented in crop plants, as well as in natural communities , over the last 50 years from phenology networks and individual records. Menzel et al. (2006), for example, report that 78 per cent of all observations in 21 European countries showed earlier flowering, with an advance in phenological events of 2.5d per decade on average. In Germany, the phenology of 78 agricultural and horticultural events between 1951 and 2004 were, on average, 1.1–1.3d earlier per decade . Likewise, winter wheat cv. Kharkof grown in the USA Great Plains has flowered 0.8–1.8d earlier per decade (depending on location) since 1950.

In addition to phenology observations, Menzel et al. (2006) also reported that farmers' activities, such as sowing and harvesting, also occurred earlier, indicating a change in crop season length. Other studies have also shown that season length has increased, at least in mid to northern latitudes, and this is associated with warmer temperatures in winter and spring. However, many of these changes in the timing of farming activities are driven by changes in farm management practices and the introduction of new cultivars. So, although several studies have associated these changes in phenology with warmer seasonal or winter/spring temperatures, earlier flowering in crop species may be related more to the earlier onset of farming activities than to temperature and hence past climate change per se. In addition, changes in crop management may also counter direct effects of temperature warming and the timing of farm operations, for example through a change to a longer duration variety. Therefore, studies that robustly attribute observed changes in phenology in ecosystems to changes in climate are rare for natural ecosystems and not found for managed ecosystems.

EFFECT OF CO_2 ON FLOWERING

The effect of [CO_2] on growth and development has been studied in many crop species. Springer and Ward (2007) recently summarized the effect of [CO_2] on flowering time in 23 crop species in 33 papers that included experiments in growth cabinets/glasshouses, open-topped chambers, and field-based FACE (free air carbon enrichment) facilities. The majority of papers compared current ambient with a doubling of [CO_2], *i.e.* the expected [CO_2] beyond 2070 depending on future greenhouse gas emissions. Effects of intermediate [CO_2] representing short- and medium-term changes in [CO_2] are not commonly reported, although changes to flowering time would be hard to detect at these intermediate [CO_2].

Approximately half the studies cited by Springer and Ward (2007) reported earlier flowering time, and only four out of 33 [Hesketh and Hellmers, 1973; Rogers et al., 1984; Ellis et al., 1995 (two species)] reported delayed flowering in response to increased [CO_2]. Earlier flowering was reported in most crop species studied, including short-day [soyabean (Glycine max), rice (Oryza sativa), and cowpea (Vigna unguiculata)] and long-day [barley (Hordeum vulgare), pea (Pisum sativum), and faba bean (Vicia faba)] species. The four papers reporting a delay in flowering were on maize, sorghum, and soyabean, all short-day species. It has been suggested that short-day and long-day species respond differently to [CO_2] (Marc and Gifford, 1984; Reekie et al., 1994). However, the four reports of delayed flowering were all from controlled-environment experiments, and there are no reports of delayed flowering from field-based experiments of short-day species.

Given that some authors have questioned the influence of artefacts associated with controlled environments on plant responses to climate , the most reliable guide to [CO_2] effects on flowering should be those reported from the FACE experiments. In FACE experiments, [CO_2] treatments are imposed on crops growing in large fields under well-managed farm conditions, *i.e.* as near 'natural' conditions as possible (Ainsworth et al., 2008). In general, the FACE experiments reported in Springer and Ward (2007), along with more recent FACE papers, suggest that [CO_2] has little or no effect on flowering time in either C4 species (*e.g.* maize; Leakey et al., 2006) or C3 species (*e.g.* rice,Shimono et al., 2009). One FACE experiment on potato (Solanum tubersosum) by Miglietta et al. (1998) does show flowering occurring 5–7d earlier at 460–660 μmol mol^{-1} than at ambient [CO_2], but this was associated with increased canopy temperature according to the paper (though no data were presented). Although reductions in stomatal conductance (g_s) are commonly reported in FACE experiments at high [CO_2], and this can increase canopy temperature, canopy temperature is not usually given.

At high [CO_2], tissue temperatures are usually increased due to lower conductance, and care is therefore required when interpreting such data. For example, Vara Prasad et al. (2006) observed tissue temperature in sorghum to be 1.2–2.7°C warmer at high than at ambient [CO_2] at ambient daytime air temperatures of 32–44°C. In this particular experiment, high [CO_2] delayed flowering slightly at 36°C compared with 32°C, suggesting a temperature $\times$ CO_2 interaction. However, this delay can be explained by higher tissue temperatures that were reported, resulting in mean temperature exceeding the optimum temperature and hence causing a delay in flowering.

TEMPERATURE AND EFFECTS ON PHENOLOGY

Given the apparent lack of a direct effect of [CO_2] on rate of development, then temperature, and interactions with temperature, will be the most important aspect of human-induced climate change for crop development. The duration from sowing to flowering and maturity in plants without a vernalization requirement, or where that requirement has been met, is largely determined by responses to temperature and photoperiod (Roberts and Summerfield, 1987; Wallace and Yan, 1998). While photoperiod-insensitive or day-neutral types are important in modern agriculture, especially in warm short-season environments, photoperiod sensitivity is the norm and is a very powerful adaptive mechanism (*e.g.* soyabean: Evans 1993; Roberts et al. 1996). The basic framework used to quantify responses to temperature and photoperiod in whole plants is given, and a close look is taken at responses above and below the optimum temperature, and tissue versus air temperature.

Phases of Development Sensitive to Temperature or Photoperiod

The rate of development of plants is generally responsive to photoperiod for only part of their life cycle (*i.e.* between emergence and flowering), though some post-flowering processes such as the rate of flower initiation are affected by photoperiod and can therefore influence the duration of the seed-filling period (Summerfield et al., 1998). In contrast, temperature affects the rate of development throughout the life cycle (Roberts and Summerfield, 1985). Three distinct stages of pre-flowering development can be identified in plants, namely the pre-inductive or juvenile, inductive, and post-inductive phases. Photoperiod only affects the duration of the inductive phase, and in photoperiod-sensitive cultivars the effects of photoperiod are usually substantial and are the major determinant of flowering time.

In most annual crop species examined, the duration of the pre-inductive or juvenile phase is short [days not weeks: soyabean ; lentil (Roberts et al., 1986)], the most important exception being in rice (O. sativa) where the duration of the pre-inductive phase (sometimes called the basic vegetative phase or BVP, though this includes part of the inductive phase) may exceed 50d . A long juvenile trait has also been identified in soyabean .

Typically the response to temperature of many plant processes, including the rate of development, is described in terms of cardinal temperatures (base or minimum, T_b; optimum, T_o; and maximum or lethal, T_m temperature) and the thermal sum (θ) or rate (inverse of the duration). This response can be quantified by simple bi-linear or broken stick models with a sharply defined T_o or by linear models with a flat-top/plateau response that have a maximum rate of development over a range of temperatures. Curvilinear quadratic or beta-function models that have a near optimal rate of development over a range of temperatures have also been used. This basic response to temperature is affected by photoperiod in photoperiod-sensitive genotypes and therefore must be determined under short-day or inductive photoperiod (Roberts and Summerfield, 1987; Yan and Wallace, 1998). These differences in model/ interpretation are not particularly important in most natural growing seasons in current climates, where mean daily temperatures are mostly close to but below the T_o; the choice of model under these circumstances makes little difference to the predicted number of days to flowering. However, in future climates where temperature is expected to exceed T_o more frequently, the choice of temperature response function will be much more significant, and this is discussed further below.

Ambient or Tissue Temperature?

Ambient air temperature is the usual 'temperature' used to quantify responses to temperature, although of course for plant processes it is the temperature nearest/approximating that of the growing point or meristem that matters . In many temperate crops, such as wheat, both soil and air temperature influence development; soil temperature while the apex is close to the ground, air temperature thereafter Similarly, in irrigated or flooded rice systems, water temperature, not air temperature, controls development until the apex is above the water (Collinson et al., 1995). It is therefore important in quantifying and modelling responses to climate change to use the appropriate 'temperature' driver, and not simply ambient air temperature. However, it is also necessary to remember that climate model output used for climate change impacts studies will provide values of surface temperature that can be viewed as similar to the

2m temperature recorded in weather stations. Therefore, quantitative relationships between the rate of development and temperature will often include a degree of uncertainty due to differences between where temperature is measured and where it is perceived by the plant.

Perhaps less obvious is that significant differences between air and tissue temperature are often found in controlled environments, and these differences may be critical for interpreting interactions where temperatures are close to the optimum. For example, Vara Prasad et al.(2000) recorded peanut (Arachis hypogeae) flower bud temperatures for plants raised in growth cabinets to be 0.4–4.3°C below ambient air temperature over the range 28–48°C. This has obvious implications for the quantification of responses to temperature for application to natural environments. Large differences between air and tissue temperatures can also arise where environmental conditions decrease transpirational cooling, for example with severe water deficits or at elevated [CO_2].

Photoperiod and Temperature Interactions

Most crop plants respond to photoperiod, and in general short- and long-day plants respond in a similar manner with photoperiods longer or shorter, respectively, than the critical or base photoperiod delaying flowering. In quantitative types, flowering is delayed but not prevented in the non-inductive photoperiod, whereas in qualitative types, if the photoperiod transgresses a critical threshold, flowering will not occur. While qualitative responses have been observed in some crop plants [*e.g.* pigeonpea ; soyabean (Roberts et al., 1996)], the photoperiod in most growing seasons does not transgress the ceiling or maximum photoperiod, or does so only for a short period, and hence most crop plants are effectively quantitative short- or long-day plants. One exception to this may be sorghum in parts of West Africa .

Most whole-plant crop models (*e.g.* DSSAT, APSIM) assume that photoperiod effects are additive to those of temperature, which is the basic underlying driver of development, though others have argued from observations that there is a photoperiod by temperature (P×T) interaction . The P×T interaction manifests itself as a hyperbolic response to photoperiod, variation in the critical photoperiod with temperature (Roberts and Summerfield, 1987), or variation in the optimum temperature with photoperiod . Sensitivity analyses, however, generally show that fixing critical photoperiods and optimum temperatures does not significantly reduce the accuracy of predictions . More recently, a threshold- or appetence-type response has been proposed wherein the threshold or target for floral initiation to occur is reduced through time . An effect of rate of change of photoperiod has also been proposed, and was included in the AFRC wheat model . As a general rule, all models predict flowering time fairly accurately in crops growing in their normal growing season where days are becoming more inductive and temperatures are favourable (March to July in temperate latitudes north of the equator; June to October in tropical/subtropical latitudes north of the equator; December to April in tropical and subtropical latitudes south of the equator); they work less well outside these norms when, for example, days are becoming less inductive (*e.g.* post-rainy season crops) or crops are planted before the longest day [*e.g.* sorghum in West Africa (Clerget et al., 2004; Dingkuhn et al., 2008)].

Most crop models assume that photoperiod only affects rate of development at and below T_0; above T_0, only temperature affects the rate. However, the effect of photoperiod at temperatures $>T_0$ has not been studied extensively and this is clearly of importance for accurately predicting phenology in future climates. Crop models such as DSSAT and APSIM effectively assume that at temperatures $>T_0$ the rate of development is only affected by temperature. In most current seasons and environments there will not be many days where temperature exceeds $T_0$and therefore this approach broadly works. However, in future climates where mean temperatures will be higher, the optimum temperature is likely to be exceeded more frequently, so interactions $>T_0$ may be more significant. This is likely to be more so in crop species such as wheat (Triticum aestivum) and common bean (Phaseolus vulgaris) that originated in temperate climates but are now widely grown in more tropical environments and which have comparatively low values for T_0 of <25°C . Crop species that originated in the tropics (*e.g.* rice, sorghum, millet, and peanut) generally have higher values for T_0, often between 25°C and 30°C, and sometimes as high as 35°C.

It is really only in controlled environments, where temperature and photoperiod can be controlled independently and supraoptimal temperatures can be applied, that these interactions can be investigated. In a series of experiments with maize (Zea mays) over several years , genotype TuxpeH´o was grown at constant temperatures and photoperiods ranging from 12°C to 36°C and 9h to 16h, respectively. The duration from sowing to tassel initiation (the first easily observed sign of reproductive apical development, cf double ridges in wheat) was recorded in 67 P×T treatments and modelled using a linear, additive regression technique assuming plants were sensitive to photoperiod from emergence (Roberts and Summerfield, 1987). The resultant modelled durations are shown as a contour graph.

Tassel initiation occurred between 18d and 111d after sowing, and the cardinal temperatures were typical for maize (T_b=6.7, T_0=28.3, and T_m=44.6°C). The genotype was sensitive to photoperiod; at 25°C, days to tassel initiation increased from 22d to 36d as the photoperiod increased from 12h to 16h, *i.e.* an increase in duration of ~60 per cent. This large and comprehensive data set suggests that photoperiod has an effect on rate of development up to and beyond T_0 in addition to any negative effect of supraoptimal temperatures.

So, for example at 30°C and 11h photoperiod, tassel initiation will occur ~20d after sowing, whereas at 15h and the same temperature tassel initiation will occur after 26d. While these effects may seem small, this delay is equivalent to a 30 per cent increase in duration to the first sign or reproductive development; were days to flowering to be modelled these delays would be much more marked.

Molecular/Genetic Aspects of Flowering Response to Climate Change

Little is known about the molecular mechanisms that control flowering times in response to ambient temperature, other than for the vernalization response, even in Arabidopsis . Similarly little is known about the response of flowering pathway genes to elevated [CO_2].

Flowering time (usually measured by leaf number) in wild-typeArabidopsis responds to temperature in a similar manner to other plants, occurring sooner at warm (up to ~27°C) than at cool (16°C) temperatures. Comparisons of mutants of the photoperiod, gibberellin, and autonomous pathways have shown that ambient temperature (growth temperature) is sensed through genetic pathways involving FCA and FVE, and integrated through FT, and that temperature leads to a photoperiod-independent activation of FT (Balasubramanian et al., 2006). Temperature also affects responses of phytochrome mutants is not able to promote flowering at cooler temperatures, a response that might contribute to P×T interaction. However, while warm/ short days (27°C/8h) apparently induce flowering at the same time as cool/long days (16°C/16h) in Arabidopsis(Balasubramanian et al., 2006), temperature cannot usually compensate for non-inductive photoperiod in most crop plants.

In Arabidopsis, Springer et al. (2008) examined the response to elevated [CO_2] in two genotypes, one selected for high seed number at elevated [CO_2] and one a random control. They found that flowering was not affected by [CO_2] in the control genotype and unsurprisingly the down-regulation of FLC (Flowering Locus C) and up-regulation of SOC1(Suppressor of Overexpression Constans 1) and LFY (Leafy) was similar at ambient and elevated [CO_2]. However, in the adapted genotype, flowering was delayed by 7–9d at elevated [CO_2] and this was associated with no down-regulation of FLC over the course of the experiment and, as a result, later up-regulation of SOC1 and LFY expression. Elevated [CO_2], as expected, increased plant size, and relative increases in the adapted and control genotypes at ambient and elevated [CO_2] were proportional to the delay in flowering. The authors suggest that higher sucrose levels may act as a signal to influence flowering. However, it may be that selection for high seed number at elevated [CO_2] was associated with later flowering and hence larger plant size and yield.

Models predicting flowering time based on flowering pathway genes have been proposed for Arabidopsis and these models do predict flowering time in constant photoperiod and temperature environments. Furthermore, these simple gene network models can generate important physiological parameters, such as critical photoperiods. However, as stated above, none of these models currently quantifies basic (suboptimal) temperature responses and certainly not supraoptimal temperatures; nor do they model temperature×photoperiod interactions.

Simulation of Phenology in Future Climate

One of the most commonly documented impacts of climate change on crops is the shortening of development stages in a warmer climate and the change in areas of crop suitability that result. In many studies this response dominates the impact of climate change on yield. Many simulation studies of changes in areas of crop suitability use simple thermal time relationships to represent the rate of development (Kenny et al., 2000; Fischer et al., 2005), and only a few model the effects of supraoptimal temperatures on the rate of development . Given that we know that photoperiod sensitivity of duration to flowering is a key determinant of crop adaptation to climate, and that many crop varieties are photoperiod sensitive to some extent, it seems important to explore how photoperiod sensitivity may affect the response of crop phenology to temperature warming. For this, the photothermal model of flowering time was used to simulate duration from sowing to tassel initiation in maize cv. Tuxpeño in the current climate at one maize-growing location and at a range of prescribed temperature increases. Specifically, the aim was to determine whether photoperiod sensitivity affected the response of duration from sowing to tassel initiation to increases in mean temperature and to a more variable temperature regime.

One hundred years of current climate at Zaria, Nigeria (11.1°N, 7.7°E), were generated using the MarkSim weather generator (Jones and Thornton, 2000). Another set of 100 years was generated with an increase in temperature variability that was simulated by increasing the diurnal temperature range by 50 per cent compared with the current climate simulations.

Temperature increases of +1, +2, to +6°C were added separately to the daily mean temperature of these two sets of current climates to provide a total of 14 sets of 100 years of climate. The photoperiod on each day was calculated from standard astronomical daylength equations . A sowing date of 15 May, the average date of the start of the rains at this location , was used in all simulations. Time to tassel initiation was simulated using the photothermal time model with the appropriate temperature and photoperiod sensitivity for maize cv. Tuxpeño (PT model), and with the same temperature sensitivity alone (T model, *i.e.* no response to photoperiod). The mean and coefficient of variation of duration from sowing to tassel initiation was calculated for each set of 100 years of simulations.

Mean duration from sowing to tassel initiation became progressively shorter with an increase in mean temperature until an apparent optimum, beyond which duration progressively increased, indicating that current mean temperatures at Zaria are below the optimum. Photoperiod sensitivity (PT model) delayed tassel initiation by 6d in the current climate (zero change in temperature) and changed the response to temperature warming. Most notably, the apparent optimum for this development stage increased from +2°C without photoperiod sensitivity (T model) to +4°C with photoperiod sensitivity. The interannual variability in the duration from sowing to tassel initiation was similar between the simulations with and without photoperiod sensitivity until +2°C, beyond which interannual variability in development was greater at a given temperature warming in the absence of photoperiod sensitivity. Also, interannual variability in development was greater when the diurnal temperature variability was increased. To explore this further, the change with temperature in the coefficient of variation of duration to tassel initiation between the climate simulations with and without extra temperature variability was examined. Again, photoperiod sensitivity affected the simulated response to temperature warming. The increase in variability of the timing of tassel initiation due to increased temperature variability at +3°C and warmer was reduced with photoperiod sensitivity.

From this preliminary analysis, it is concluded that photoperiod sensitivity changed the response of flowering time to simulated temperature warming that is a typical component of a climate change impacts study. Specifically, the apparent optimum warming for rate of development is warmer, and the increase in the variability of flowering time due to a more variable temperature is less, when the rate of development to flowering is sensitive to photoperiod and temperature, compared with temperature alone. An implication of these results, although they are based on only a single set of simulations for one genotype of maize at one location, is that climate change impact studies that only use thermal time to model crop development may not be capturing the correct response of development to climate change for crop genotypes that are photoperiod sensitive. Given the importance of changes in development in a warmer climate for crop yield and the timing of crop-sensitive phases, this is clearly a topic that merits further research.

CANNABIS PLANTS - CULTIVATION AND YIELDS

CLASSIFICATION AND OVERVIEW

Authorities disagree about the number of species of plant which constitute the genus Cannabis. Although many authorities continue to class all varieties of the plant, including Hemp and Marijuana, as Cannabis sativa, it is widely accepted that there are three separate species or sub-species: Cannabis sativa being most widely cultivated in the Western World, was originally grown on an industrial scale for fibre, oil, and animal feedstuffs, is characterised by tall growth with few, widely-spaced, branches; Cannabis indica, originating in south Asia, and also known historically as Indian Hemp, was cultivated for the drug content, with shorter bushy plants giving a much greater yield per unit height; Cannabis ruderalisis a hardier variety grown in the northern Himalayas and southern states of the former Soviet Union, having a more sparse "weedy" growth, and is rarely cultivated for the drug content. Those who argue that all three are one species point to the fact that cross-breeding produces viable and fertile daughter plants.

Cannabis is unusual among plants in that it is dioecious, *i.e.* any one plant may be either male or female. Most everyday flowering plants are hermaphroditic, *i.e.* flowers contain both stamens (male) and pistils (female), whereas the cannabis plant normally produces flowers of one sex or the other. Although the main psychoactive compound - D^9 tetrahydrocannabinol, or THC - is present in leaf and in male plants and flowers, the concentration is usually greatest in the female flowering tops. In most cultivations, plants are produced from cuttings taken from known female plants.

Cannabis grown for drug content is normally grown under controlled conditions, where male and female plants are separated, allowing the female flowers to grow without producing seed (sinsemilla), and produce the most potent material. Where cannabis plants are propagated from cuttings it is possible to determine the sex of the plant so that only females are grown. Where plants are grown from seed, 50 per cent of plants should be of either sex, although some seed catalogues have claimed to guarantee female plants. These are possible due to the tendency of some female plants to produce a few male flowers (hermaphrodites), enabling pollination with female genetic material, although the plants resulting from hermaphroditic seeds may be unpredictable, unhealthy, and inbred (especially where cuttings originate from the same mother plant). The whole of the cannabis plant, with the exception of the seeds and mature stalk, are prohibited under the Misuse of Drugs Act 1971. However, it is only the dried tops of female flowers which are of value, being the form in which herbal cannabis is normally sold. This material is clearly distinguishable from cannabis leaf, which is not regarded as of "merchantable quality", and as such should only be allotted a nominal value.

The quality and yield of cannabis is governed by two factors, the genetic makeup of the seedstock, and the environment in which the cannabis is grown.

Genetic Factors:

Cannabis can be grown from a variety of genetic sources. The parent stock of the plant determines factors such as growth rate (size), internodal length (bushiness), flowering time, flower colour and development and THC content.

Hemp seed - sold in pet food shops. This is the bottom end of the quality spectrum, producing very tall leafy plants which physically resemble cannabis but have virtually no THC content (<0.3 per cent) Although seeds are irradiated in order to cause sterility, a percentage will flower and may produce plants of substantial size, with appearance similar to other cannabis plants. The THC content will be very low, normally less than 0.3 per cent. Industrial hemp can be cultivated under licence in the UK.

Seeds from herbal cannabis deals. The cannabis grown will be similar to the parent crop, but mature seeds are themselves an indication of poorer-quality cannabis (*i.e.* not sinsemilla)

Pedigree Seeds - Many companies and internet sites offer "pedigree" seeds to be bought in order to grow cannabis of high quality. There are a wide numbers of cultivars available, from more traditional 'Skunk, Haze, Big Bud, Northern Lights' to newer varieties such as 'White Widow, Bubblegum, Blue Cheese, Jack Herer, Mr Nice, AK47' and many more.

Cuttings - As seeds for "pedigree" plants can be expensive, plants are frequently propagated from cuttings, making clones of the "mother" plant identical in genetic makeup guaranteeing a female plant. If flowering is prevented, a single plant could produce an unlimited amount of material if cuttings were continuously taken and grown from "mother, daughter, granddaughter (etc.)" plants. There is a limited trade in cuttings between growers.

Breeding - Growers can produce their own seeds from first generation crops, these are sometimes used for breeding purposes. In such cases male plants would be cultivated in a separate isolated system.

Environmental Factors:

These include the temperature, soil composition, supply of nutrients and water, the spectrum of light available for photosynthesis and the length of the daylight period. The potential yield is primarily limited by the available space, and also by overfeeding, overwatering, disease or infestation.

- *Temperature:* Cannabis grows most efficiently at summer temperatures of 20° to 25°C. Higher temperatures are tolerated for short periods, but lower temperatures reduce growth and resin development. Although cannabis grows well outdoors in the UK, autumn temperatures are generally too low for optimum flower development and resin production. Indoors, temperature is commonly regulated using thermostatically controlled fans and heaters.
- *Atmosphere:* Indoors it is normal to find oscillating fans of various sizes to circulate air amongst the plants and to stress the stems leading to stronger growth. Extractor fans are very common, usually with ducting to take the stale air to the exterior of the building, and

creating negative pressure drawing fresh air into the system and minimizing escape of odours. Air filters are increasingly common to reduce the characteristic smell of cannabis in the vicinity of grow-rooms Depletion of atmospheric CO_2 by photosynthesis of plants grown in confined spaces may be countered by proper ventilation, or by additional CO_2 from commercial cylinders, burning fossil fuels (*e.g.* gas heaters) or occasionally from home-brewing beers, ciders or wines in the same area.

- *Growth Media:* Growers use a variety of growing media from plant pots/buckets and growbags with conventional soil and compost types, to hydroponic systems. In hydroponic systems water and nutrients are controlled via a continuous circulation to plants grown without soil in tanks using inert media such as gels, rockwool, clay granules or other materials which retain water and stimulate root growth. Cannabis grows most effectively with a neutral to slightly alkaline pH.
- *Lighting:* Growers can employ a variety of light sources, from the sun to artificial light. Availability of light is a key determinant of yield, as low light intensities produce spindly growth and inhibit development. Artificial sources include

Fluorescent tubes of the Gro-lux type used in aquaria, these produce the correct spectrum for maximum photosynthesis but are of low intensity. Similar tubes are commonly used for propagation of cuttings, often in purpose-built arrays. Low-Energy Grow-Lights – these are like the household low-energy bulbs but with higher outputs, drawing between 100 and 250 watts. These are becoming increasingly common.

High-intensity metal halide and mercury vapour bulbs give the blue-white light spectrum which induces maximum vegetative growth. High pressure sodium (HPS) lights give a more reddish/yellow spectrum and are considered to produce more abundant flower growth. These come in sizes from 250W to 1000W, most commonly 600W.

A growing number of UK retailers are supplying such lighting systems suitable for the small indoor grower and the prices commanded by such lights are falling. There is a limited second-hand market, but industrial lights and fittings often use similar bulbs which may be available at public auctions. HPS bulbs are even found lighting some court buildings, and are increasingly used for street lighting in preference to the low-pressure yellow sodium vapour lights.

- *Photoperiod:* The length of the daylight period controls the flowering of the plant. Long days, of up to 18 hours sunlight, cause the plant to grow in size, or vegetatively. For the plant to flower, the day length needs to be 10 to 12 hours (in actual fact it is the night or dark period which must be 12 to 14 hours), to make the plant 'believe' that winter is approaching. Outdoors in the UK, the day-lengths in autumn shorten more rapidly the further north the site is located, falling from 12-13 hours to below 10 hours daylight within a few weeks. This reduces the duration of the flowering season, and so the proportion of flowering material in outdoor plants (including those greenhouses) will be much lower than in indoor systems with a constant 12-hour cycle.

- *Separate Rooms:* Most indoor growers will have more than one growing room. One room will have a day length of 24 or 18 hours, and often use a blue-white light source for optimum vegetative growth. The other room (or rooms) will have a 12 hour day length with blackout conditions if necessary, and frequently use a redder HPS light to induce maximum flower development. It is also common for cuttings to be propagated separately from the main growth area, under fluorescent or lower powered lights.
- *Available Space:* The available space is the clearest limitation on potential yield of cannabis plants if these are grown indoors. The yield of plants at maturity is determined by the height and lateral growth, both of which may be restricted in an indoor set-up.

Vertical Space – the maximum height a plant can achieve is the distance from the base of the stalk (*i.e.* top of pot or level of tray top in hydroponic systems) to around 15cm beneath the lighting canopy. In practice, plants are usually flowered around 30-60cm beneath the lighting. Once plants are flowering little increase in height is expected. Shorter plants will normally yield less than taller plants of the same variety grown in the same conditions.

Lateral Space – the taller the plant the greater the scope for side-branching, if the spacing between plants is less than around half the plant height the development of side-branches will be restricted. The degree of branching is also dependent on the internodal length. Clearly, where a large number of plants are packed into a small space the yield from each plant will be substantially reduced. Indeed, the overall yield from 50 plants may not be significantly different from the overall yield from 10 plants grown in the same space.

Attrition and Wastage:

In any system, the limiting factor on yield is the number of plants which can be grown in the flowering stage at any one time. Where a grower has a large number of seedlings and/or cuttings it is not reasonable to assume that all of these will be grown to maturity.

- *Poor stock:* A proportion of plants grown from seed will die or be unsuitable, through abnormalities in growth and development.
- *Failure to root:* A proportion of cuttings will fail to "take" *i.e.* develop roots in order to grow independently. Growers will frequently select

the healthiest plants for further growth and development, and discard plants which are weedier or underdeveloped.

- *Male plants:* Where plants have not been sexed, up to 50 per cent of seedlings/cuttings, yielding male flowers, will be discarded at an early stage in order to prevent pollination of female flowers.
- *Poor conditions:* The yield of a crop may be substantially reduced by extremes of temperature and humidity, by over-watering or over-fertilising which can permanently damage the plants..
- *Disease/Infestation:* A proportion of plants may die or become weakened, attacked by insects or plant diseases. Several plant diseases can attack cannabis, which in an indoor garden can rapidly destroy virtually the entire crop. Insect pests such as spider mites and aphids (greenfly) are frequently encountered.

Cannabis Yields – Literature Sources

Studies of cannabis plants grown in the UK under outdoor conditions by the Laboratory of the Government Chemist have shown a range of dry weights (gross) per plant of 10.9 to 59.1g , 16 to 106g and 8 to 80 g , with mean weights per plant of 30 to 60g. The variations in growth rates and THC content were ascribed largely to weather conditions. Other sources have stressed the variability of yields and the importance of environmental factors. Kaa studied plants grown outdoors and in greenhouses in Denmark finding median gross weights of 308g and 584g respectively, with a mean yield of 8.7 per cent flowering tops after drying. GW Pharmaceuticals, who have a Home Office Licence to grow cannabis in the UK, report gross yields of 157g-188g m^{-2} in greenhouse conditions, 251g-397g m^{-2} indoors under mercury lights, and 516g-573g m^{-2} under HPS lighting (at between 10 and 17× plants per square metre.

For indoor cannabis cultivation, Ed Rosenthal, author of a number of books on cannabis cultivation, in evidence to the U.S. Congressional Sentencing Commission, stated that a mature cannabis plant grown under modern indoor conditions can usually be expected to yield 10 grams of marijuana (*i.e.* dried flowering tops), and that each "marijuana garden" should be treated on its own merits. Knight et al grew plants hydroponically under optimum conditions with mesh support for branches ('screen of green') yielding a mammoth average yield of 687g per plant. In more typical hydroponic growing conditions in the Netherlands Toonen et al reported an average yield of 33.78g per plant, and Huizer et al reported an average 22g per plant.

Forensic Evidence and Databases:

In the UK, the Forensic Science Service and Environmental Scientifics Ltd maintain separate databases of seized cannabis plants, with the FSS currently (2010) reporting an average yield of 40g per plant and Scientifics between 50g and 58g per plant (although their 10 per cent trimmed mean was recently quoted at a more reasonable 46.5g. Neither of these make any allowance for height, cultivation method or available space, and their figures are dependent on officers submitting representative plants from cultivation systems, rather than the largest plants present. Previously the FSS were reporting an average 10-15g per plant. A further firm, LGC Forensics, estimates yields from immature plants on the basis of 14g per foot height. The equivalent IDMU figures would range between 7g per foot for long-internode plants to 18g per foot for short-internode plants, and 12g per foot for medium-internode plants. The LGC estimate thus appears a more reasonable basis for estimation than a crude average which takes no account of the size of plants.

In 1998 FSS scientists Bone and Waldron reported "The amount of usable cannabis recovered per plant varies enormously, depending on the bushiness of the plant. Weights recovered have been found to range from 1 to 355 grams per plant. In the more mature plants, flowering head material may be 50-70 per cent of this weight."

Yields of Cannabis Plants in Previous IDMU Cases

IDMU staff have been acting as expert witnesses in criminal cases involving cannabis plants since 1991 where standard procedures include examination of seized cannabis plants at police stations or forensic laboratories. Since 1994 routine data collection included measurement of heights, internodal lengths (maximum on main stem), and yields of leaf and flowering tops. To January 2011 the database included 2409x plants of which 1478 were 'mature' and 931 were immature.

Table. Mature Cannabis Plant Yields in IDMU Cases (1994-2010).

Height	No of	Yield of Flowering Tops			% Yield	Yield by Internodal Length		
	Plants	Min	Max	Average	Tops	Short	Medium	Long
<25cm	135	0.5g	14.8g	3.11g	66.5%	3.24g	4.54g	2.20g
26-40cm	207	0.7g	30.1g	11.69g	62.0%	16.24g	8.51g	12.45g
41-60cm	229	2.12g	118g	16.25g	65.8%	31.78g	19.15g	5.25g
61-80cm	303	1.58g	154g	27.32g	69.2%	66.36g	34.50g	10.16g
81-100cm	307	2.80g	126g	32.04g	66.4%	46.34g	47.14g	22.05g
101-120cm	156	2.54g	221g	34.65g	66.7%	53.04g	33.43g	27.39g
121cm +	141	11.04g	298g	41.55g	70.4%	68.00g	37.08g	38.53g
Overall	1478			24.32g	64.2%	595mg	393mg	211mg

Mature Plants: The IDMU database includes a total of 1478x mature (i,e, over 50 per cent flowering tops) sample and bulk flowering plants, found a range between 0.5g and 298g flowering tops (average 24.3g) from individual mature plants with a height range of 8cm to 2.1 metres. The average proportion of flowering tops (excluding figures over 95 per cent which represent stripped plants) was 64.2 per cent. Heights and maximum internodal lengths (short 4cm or less, medium 4.1-7.9cm, long 8cm plus) were also recorded, and results are shown in table below:

Clearly the single most important factor in predicting yield is the height of plants, although the relationship is not linear. For a given height, plants with short internodal lengths tend to produce higher yields (595mg/cm) than those with medium (393mg/cm) or long internodes (211mg/cm), although a heavily-branched plant with long internodes will produce a higher yield than a single-stemmed plant with short-medium internodes.

Immature Plants: A total of 931x immature flowering plants were also recorded (defined as comprising 49 per cent flowering tops or less by weight). Extrapolating the yield of leaf material at seizure by a factor of 1.79 (allowing for average 64.2 per cent tops at maturity) gives good correspondence with actual yield figures from mature plants for the corresponding height ranges ('estimate' maturity comparison values are compared with mean and median values from the mature plants of corresponding heights).

Table. Immature Cannabis Plant Yields in IDMU Cases (1994-2010).

Height	No of	Average Yields at Seizure			% Yield	Maturity Prediction	
	Plants	Leaf	Tops	Total	Tops	Estimate	Actual
<25cm	*186*	2.54g	1.14g	3.68g	31.0%	4.55g	3.11g
26-40cm	*89*	13.66g	6.03g	19.69g	30.6%	24.45g	11.69g
41-60cm	*218*	11.46g	5.82g	17.28g	33.7%	20.51g	16.25g
61-80cm	*187*	15.43g	4.97g	20.38g	24.4%	27.62g	27.32g
81-100cm	*101*	23.80g	10.14g	33.94g	29.9%	42.60g	32.04g
101-120cm	*113*	16.18g	5.16g	21.23g	24.3%	28.96g	34.65g
121cm +	*37*	38.58g	16.78g	55.36g	30.3%	60.06g	41.55g
Total	*931*	13.67g	5.56g	19.23g	100.0%	24.5g	24.3g

Hydroponic vs organic systems: Where plants are grown hydroponically (in rockwool or clay granules) rather than organically (in compost/soil media), yields tend to be similar with short-medium height plants but higher for taller plants in the range 80-120cm. Yields from plants grown outdoors or in greenhouses are limited by the rapid decrease in day length during autumn, with the highest recorded proportion of flowering tops in outdoor-grown plants being 44 per cent.

Table. Yield by Height and Growth Method.

Method	Hydroponic			Organic			Outdoor		
Height	n	Tops	Total	n	Tops	Total	n	Tops	Total
<25cm	90	3.50g	5.30g	43	2.28g	3.27g	2	4.90g	14.00g
25-40cm	190	12.06g	19.55g	17	7.56g	11.27g	2	2.22g	10.10g
40-60cm	123	14.30g	22.13g	102	18.76g	28.14g	4	12.08g	17.70g
60-80cm	166	28.77g	44.45g	135	25.90g	33.90g	2	4.10g	11.70g
80-100cm	87	52.87g	80.81g	218	23.68g	35.03g	15	5.91g	21.73g
100-120cm	59	47.06g	68.92g	93	26.20g	39.34g	14	9.73g	29.70g
>120cm	81	38.05g	52.57g	57	44.67g	63.22g	12	28.24g	85.22g
Total	796	24.62g	37.24g	665	23.73g	34.10g	51	12.44g	37.39g

CALCULATION OF YIELDS

Mature Plants: Where cannabis plants are mature, potential yields can be calculated from yields of sample plants, where sample plants are representative of the crop as a whole. The yield is calculated as the average yield of flowering tops per plant multiplied by the number of flowering plants in the system.

Immature Flowering Plants: Where an intact cannabis plant is flowering but not fully mature the yield at maturity can be predicted using the following reasonable presumptions.

a. Once significant flower clusters appear (about 2-3 weeks into the flowering cycle) leaf production effectively ceases and the plant ceases growing in height
b. At maturity the average proportion of flowering tops is known.

The predicted yield of flowering tops can therefore be calculated by multiplying the leaf yield at seizure by the ratio of flowering tops to leaf at maturity, *i.e.* 1.79 (179 per cent)

Immature vegetative plants: Where no plants are found in the flowering stage a minimum yield can be predicted based on average plant yields.

PHOTORECEPTORS AND REGULATION OF FLOWERING TIME

One of the most important environmental factors affecting flowering time is the daily duration of light, the photoperiod, which was first discovered by Garner and Allard in the 1920s (Thomas and Vince-Prue, 1997, and refs. therein).

Plants in which flowering occurs or is accelerated in short days (SD) or long days (LD) are known as SD plants or LD plants, respectively. LD plants often flower in later spring or early summer (when the daylength becomes longer) to set seeds in a favourable season. SD plants generally flower in fall (when photoperiods are getting shorter) to finish reproduction before the cold winter arrives. Synchronization of flowering time with a reliable environmental cue such as the photoperiod also increases the chance of out-breeding and genetic recombination.

The photoperiodic control of flowering is brought about by the interactions of genes involved in the developmental control of floral initiation, the regulation of the circadian clock, and the signal transduction of photoreceptors . Recent molecular genetic studies in a facultative LD plant, Arabidopsis, have made notable progress in identifying genetic pathways and molecular components associated with the control of flowering time and the function of the circadian clock, which have been discussed in two recent Updates(Pineiro and Coupland, 1998; Somers, 1999).

This Update focuses on the recent advances in our understanding of plant photoreceptors phytochromes and cryptochromes, and their roles in the regulation of flowering time.

CONTROL OF PLANT FLOWERING TIME

Genetic Pathways Control Flowering Time

Flower formation is initiated by the transition of the apical meristem from a vegetative fate to a floral fate. Mechanisms that control the timing of floral initiation have been extensively studied in Arabidopsis by the identification of mutations that flower earlier or later than the wild type but otherwise remain healthy (Koornneef et al., 1998).

These mutations are known as flowering-time mutations and the corresponding genes are known as flowering-time genes. In addition, many genes that were initially studied for their roles in other aspects of plant development, such as light perception, hormone metabolism, signal transduction, and floral meristem specification, also play roles in the regulation of flowering time and are sometimes also referred to as flowering-time genes. Based on phenotypic and genetic epistasis analysis of these mutations, flowering-time genes have been grouped into several signal transduction pathways that either suppress or promote floral initiation. These signaling pathways transmit either the developmental or environmental signals to regulate the expression of the floral-meristem-identity genes that control the formation of the floral meristem. Readers are referred to two recent reviews for detailed discussions of genes associated with these pathways in Arabidopsis (Koornneef et al., 1998;Levy and Dean, 1998).

Genes of the Photoperiodic Pathway

One of the major signal transduction pathways regulating flowering time is known as either the LD promotion pathway (Koornneef et al., 1998) or the photoperiodic pathway (Levy and Dean, 1998), which relays light and photoperiodic timing signals to the floral initiation process. Mutations of genes in this pathway reduce a plant's responsiveness to photoperiods. As a facultative LD plant, Arabidopsis grown in LD conditions flowers earlier than when grown withinSD. Misexpression of genes associated with the LD pathway may also delay the flowering of Arabidopsis plants grown in LD, but does not alter the flowering time of plants grown inSD, resulting in reduced sensitivity (hyposensitive) to photoperiod. Mutations in genes such as CO(CONSTANS; Putterill et al., 1995),PHYA (phytochrome A; Johnson et al., 1994; Reed et al., 1994), CRY2 (cryptochrome 2; Guo et al., 1998), andGI(GIGANTEA; Fowler et al., 1999; Park et al., 1999), are of this type. The elevated expression of the CCA1(circadian clock associated; Wang and Tobin, 1998) andLHY(late elongated hypocotyl; Schaffer et al., 1998) genes also results in photoperiod-hyposensitive late-flowering.

On the other hand, a mutant that flowers earlier than the wild type in both LD andSD may also have reduced sensitivity to photoperiod. Early-flowering mutations in genes such as PHYB(phytochrome B; Goto et al., 1991), PHYD, PHYE, ELF3, andPEF belong to this group. We often assume that a late-flowering mutation corresponds to a gene product that normally promotes floral initiation, whereas an early-flowering mutant implies that the corresponding gene product is a suppressor of floral initiation. Not surprisingly, many genes isolated to date that are associated with the photoperiodic pathway encode either photoreceptors or proteins associated with the circadian clock.

PLANT PHOTORECEPTORS

The primary photosensory receptors of higher plants are the red/far-red light receptors called phytochromes and the blue/UV-A light receptors called cryptochromes (Kendrick and Kronenberg, 1994). Blue light (approximately 400–500 nm) and red light (approximately 600–700 nm) are the two spectra of solar radiation that are most effectively absorbed and utilized by the photosynthetic system of plants. Therefore, the regulation of plant development by phytochromes and cryptochromes allows plants to optimize their developmental processes in coordination with the availability of energy and metabolite resources.

Phytochromes

Phytochromes are photochromic proteins that exist as two photo-interconvertible isomeric forms: the red-light-absorbing form (Pr) and the far-red-light-absorbing form (Pfr; Kendrick and Kronenberg, 1994; Hughes, 1999).

Arabidopsis has five phytochrome genes, PHYA to PHYE, which encode the apoproteins of PHYA to PHYE, respectively (Quail et al., 1995). Mutations in four of the Arabidopsis phytochrome genes have been isolated and studied. Different phytochromes regulate either distinct light responses or similar responses under different light conditions (light quantity, quality, and timing). Taking the well-characterized light-inhibition of hypocotyl elongation as an example (Quail et al., 1995), the phyA mutant is impaired in hypocotyl inhibition in far-red light, but not in red light. Conversely, the phyBmutant loses the ability to inhibit hypocotyl elongation in red light, but not in far-red light, suggesting that although phyA and phyB both mediate light inhibition of hypocotyl elongation, phyA functions primarily in far-red light, whereas phyB acts mainly in red light.

Cryptochromes

Cryptochromes are flavoproteins that share amino acid sequence similarity with DNA photolyases that catalyze blue/UV-A light-dependent DNA repairing (Sancar, 1994; Cashmore et al., 1999). Cryptochromes have no DNA photolyase activity; they usually have a C-terminal domain with little sequence homology to photolyase and they show characteristics of blue/UV-A light receptors in plants. Arabidopsis has at least two cryptochrome genes, CRY1 and CRY2. Similar to phytochromes, genetic studies of Arabidopsis cryptochrome mutations affecting light-dependent hypocotyl inhibition have played a critical role in our understanding of cryptochromes. The isolation of an Arabidopsis mutant, hy4, which has an elongated hypocotyl in blue light, allowed the cloning of the first cryptochrome gene (Koornneef et al., 1980; Ahmad and Cashmore, 1993).

The HY4 gene, later referred to as CRY1, encodes a protein associated with a flavin chromophore (FAD) that absorbs blue/UV-A light, as was previously suspected for a plant blue/UV-A light receptor . A plant's sensitivity to blue/ UV-A light can be altered by changing the expression levels of cry1 . The second Arabidopsis cryptochrome gene, CRY2, was cloned usingCRY1 cDNA as the hybridization probe . The amino acid sequence of CRY2 and CRY1 are about 50 per cent identical, but most of the sequence similarity is concentrated in the N-terminal photolyase-like domain, whereas the C-terminal domains are quite diverged . Interestingly, cry2 protein is rapidly degraded in etiolated seedlings exposed to blue light (Lin et al., 1998; Guo et al., 1999), which is reminiscent of the red-light-induced degradation of phyA (Clough et al., 1999, and refs. therein). It is not clear what functional role the light-induced proteolysis of phyA and cry2 may play, but no diurnal change in the protein expression levels has been reported for cry2.

Based on the observation that transgenic plants overexpressingCRY2 were hypersensitive to blue light, a genetic screen was designed to look for additional

Arabidopsis mutants exhibiting a long hypocotyl in blue light. Surprisingly, the resulting cry2 mutants derived from this screen showed a more apparent abnormality in flowering time than in hypocotyl inhibition and turned out to be allelic to fha, a photoperiod-hyposensitive late-flowering mutation previously characterized by Koornneef.

Since the isolation of the Arabidopsis CRYs, cryptochrome genes have been isolated from not only other plant species and algae, but also animals including fruit fly, mouse, and human. Studies of mouse and fruit fly cryptochromes have indicated that these proteins play important roles in the function and regulation of animal circadian clocks.

HOW PHOTORECEPTORS WORK

How do photoreceptors convey light signals to affect cellular processes? What are the early steps of photoreceptor signal transduction? A photoreceptor may relay light signals to other molecules by a light-dependent enzymatic activity, or it may do so by changing its conformation and thus its interaction with signaling partners. It appears that at least for phytochromes, the early steps of the signal transduction involve both types of reactions: phytochromes are protein kinases and can interact with signal transducing proteins in a light-dependent manner.

Phytochrome Kinases and a Phytochrome-Regulated Kinase

Plant phytochromes were first proposed to be protein kinases more than a decade ago , but this has remained a controversial proposition until recently. In light of concurring evidence, the phytochrome kinase hypothesis has gradually gained general acceptance.

A pivotal question that has arisen from this decade-long debate is the identification of the bona fide substrates of the phytochrome kinases concerning photo-signal transduction in flowering plants. Two recent reports have directly addressed this question. One contender for the substrate of phytochrome kinases turned out to be the newly discovered cryptochrome. It was reported that cryptochromes can interact with phyA in vitro, and in the yeast two-hybrid assay, the recombinant CRY1 could be phosphorylated in vitro by recombinant oat phyA protein. Also, the in vitro phosphorylation of cry1 by phyA was more efficient in red light or blue light than in the dark. The phosphorylation of cry1 was also found to occur in vivo in a red-light-dependent and far-red-light-reversible manner, which again suggested the involvement of a phytochrome. However, the physiological relevance of phytochrome-dependent phosphorylation of cry1 remains unclear.

Another possible substrate for phytochrome kinase is PKS1. The gene encoding PKS1 was isolated from a yeast two-hybrid screen using the C-terminal domain of Arabidopsis PHYA as the "bait." PKS1 can be phosphorylated in vitro by the recombinant oat phyA. Although PKS1 binds to both the Pr and Pfr forms of phyA, Pfr is a more active kinase than Pr in the phosphorylation of PKS1. In keeping with PKS1 being a substrate of phytochromes, PKS1 was phosphorylated in vivo in a red-light-dependent manner. The observation that PKS1 was hyperphosphorylated in transgenic plants overexpressing phyB suggested an involvement of phyB in the phosphorylation of PKS1 in vivo. PKS1 played a negative role in phyB signaling, because transgenic plants overexpressing PKS1 showed the phenotype similar to that of a phyBmutant: transgenic plants overexpressing PKS1 had elongated hypocotyls in red light but not in blue or far-red light .

Phytochrome may also regulate the activity of other protein kinases. For example, a recently identified phyA-interacting protein, nucleotide diphosphate kinase 2 (NDPK2), appears to be such an enzyme . In an in vitro binding assay, the Pfr form of phyA could bind to NDPK2 about three to four times better than the Pr form. The binding of Pfr (but not Pr) to NDPK2 increased the substrate affinity of this kinase in an in vitro NDPK2 enzymatic assay. NDPK2, localized in both the cytosol and nucleus, may play a positive role in phytochrome signal transduction. An Arabidopsis mutant with theNDPK2 gene interrupted by a T-DNA insertion showed decreased sensitivity to both red light and far-red light in cotyledon opening and greening .

Plant Photoreceptors Can Enter the Nucleus and Interact with Nuclear Proteins

Where do photoreceptors work in the cell? Phytochromes and cryptochromes are both soluble proteins, and it seems clear now that both types of photoreceptors can enter the nucleus, either constitutively or in a light-dependent manner. The intracellular localization of phytochromes and cryptochromes has been studied using fusion protein assays. In these studies, a transgene encoding a fusion protein of a photoreceptor and a marker enzyme such as β-glucuronidase or green fluorescence protein is expressed in plants, and the intracellular localization of the photoreceptor is identified by monitoring the location of the visible marker enzyme. These studies demonstrated that phyA and phyB stay mostly in the cytosol in the dark, but are translocated to the nucleus in the light. Arabidopsis cry1 and cry2 are also nuclear proteins, although no light regulation of the nuclear transportation of cryptochromes has been reported.

Differential nuclear compartmentation is commonly found for receptor molecules in eukaryotes (Adam, 1999). In the absence of a ligand, a receptor can be bound to a cytosolic protein and thus be retained in the cytosol. Interaction with the ligand may induce the translocation of the receptor to the nucleus. A similar sport may also be played by plant photoreceptors. For example, it has been suggested that PKS1 may act as a cytosolic-retention protein for phytochromes in non-inductive conditions such as the dark. Light induces the nuclear compartmentation of the phytochrome. Once in the nucleus, a photoreceptor may interact with other nuclear proteins to affect light-regulated gene expression (Smith, 1999).

Recently, two phytochrome-signaling nuclear proteins, SPA1 and PIF3, have been identified. The SPA1 gene was identified by positional cloning of the spa1(suppressor of phyA-105) mutation that was isolated as an allele-specific suppressor of aphyA mutation (Hoecker et al., 1998, 1999). SPA1 is a nuclear protein that has WD repeats, a coil-coil domain, and a protein kinase domain. spa1 mutant plants exhibited an exaggerated hypocotyl inhibition in response to light, suggesting that SPA1 is a negative regulator of phyA. The expression of SPA1 was up-regulated in light through the action of phytochromes, indicating a possible feedback regulation on the phytochrome signal transduction.

The gene encoding another phytochrome-signaling nuclear protein, PIF3 (phytochrome interacting factor), was isolated on the basis of its interaction with the C-terminal domain of PHYB in a yeast two-hybrid assay . PIF3 contains a PAS protein-protein interaction domain and a basic helix-loop-helix (bHLH) domain that may have a role in the interaction with promoters of light-regulated genes. The in vitro interaction between PIF3 and phytochromes was dependent on red light: PIF3 interacted strongly with the Pfr form, but only weakly with the Pr form of Arabidopsis phyB .

In contrast to SPA1, the expression of PIF3 was down-regulated in light . PIF3 is a positive regulator of phytochrome function, because PIF3-antisense transgenic plants showed a reduced hypocotyl inhibition in response to light. Consistent with the hypothesis that nuclear compartmentation and binding of phytochrome to a nuclear protein such as PIF3 may be part of the signal transduction leading to the regulation of gene expression, PIF3-antisense transgenic plants exhibited reduced light responsiveness for expression of various light-regulated genes .

It is interesting that although every Arabidopsis photoreceptor studied to date has been shown to play a role in both light-regulated hypocotyl inhibition and floral initiation, most mutations or transgenic plants misexpressing the phytochrome-signaling genes described above, except PIF3, have no reported alteration in flowering time. Since some of these phytochrome-signaling factors bind to phytochrome, it may be argued that there are two separate phytochrome signal transduction pathways leading to the two different developmental responses. However, some of the flowering-time genes associated with the photoperiodic pathway (*e.g.* ELF3,CCA1, LHY, COP1, and DET1) have functions in both hypocotyl inhibition and flowering time. These observations may not be satisfactorily explained by models involving linear signal transduction pathways of photoreceptors, even though we frequently use the term "pathway" in this Update.

HOW INDIVIDUAL PHOTORECEPTORS REGULATE FLOWERING TIME

In searching for photoreceptors regulating photoperiodic responses, action spectra have been extensively analyzed in different plants to investigate how light qualities affect flowering time. These early studies, along with observations of the effect of light on germination and stem elongation, led to the discovery of phytochrome. The more recent studies of photoreceptor mutations have allowed us to assign specific functions of individual photoreceptors in the regulation of flowering time.

Effect of Light Quality on Arabidopsis Flowering Time

Two of the most frequently used experimental approaches to analyze the action spectra of light regulation of flowering time are the day extension and the night break methods . In a day extension experiment, the SD photoperiod is extended by applying a low-fluence-rate light at the end of the main photoperiod. For a night break experiment, the additional light exposure is often inserted in the middle of a long night. Both conditions mimic the LD photoperiod that promotes flowering in LD plants, and are referred to as "quasi LD" in the following discussion. In Arabidopsis, far-red light, blue light, and red light were all effective at promoting flowering in night break experiments, although red light was the least effective. Day extensions with far-red light or light rich in far-red spectra (*e.g.* incandescent light) are also very effective in promoting flowering.

Fig. Effect of Red and Blue Light on Arabidopsis Flowering Time. The 21-d-old Arabidopsis (Ecotype Columbia) Plants Shown were Imbibed at 4°C for 4 d in the Dark and then Grown Under Continuous Blue, Red, or White Light (All Approximately 100 $\mu mol\ s^{-1}m^{-2}$). The Plant Grown in Blue Light had Flowered for 6 d, that in White Light for 1 d, and the Plant Grown in Red Light was not Flowering (it was about 20 d Before Flowering).

Continuous illumination with light of different wavelengths is another method used to investigate how different photoreceptors regulate floral initiation. Although the night-break and day-extension methods can more effectively minimize the interference of photosynthesis than the continuous-light methods, the latter condition can simplify the situation by eliminating light/dark cycles (and the influence of the circadian clock) to allow an assessment of the direct effect of photoreceptors on floral initiation.

Arabidopsis plants grown under continuous light with a high red- to far-red-light ratio flower later than plants grown in light of a low red- to far-red-light ratio (*i.e.* rich in far-red light;Halliday et al., 1994). Moreover, plants grown in continuous red light flower significantly later than those grown in continuous blue light. Therefore, the rule of thumb seems to be that, at least for Arabidopsis, far-red light and blue light promote flowering, whereas red light is often inhibitory.

phyA

phyA promotes flowering. The Arabidopsis phyA mutant flowers later than wild-type plants in LD or quasi-LD conditions with either night breaks or day extensions. Consistent with the notion that phyA plays a promotive role in flowering, transgenic Arabidopsis plants overexpressing phyA flowered earlier than the wild type in both SD and quasi-LD conditions. Both phyA mutant plants and phyA-overexpressing transgenic plants had decreased sensitivity to photoperiod, because they flowered at about the same time in SD as in the quasi-LD conditions. ThephyA mutant of pea, another LD plant, also showed a phenotype similar to that of the Arabidopsis phyA mutant. The pea phyA mutant (fun1) flowered normally inSD photoperiods but failed to respond to day-extension treatments with incandescent light; therefore, the peaphyA mutant flowered at about the same time inSD and quasi-LD conditions . Interestingly, the pea phyA mutant accumulated a graft-transmittable inhibitor that could delay the flowering of the grafting recipient plants, suggesting that phyA signaling may suppress the biosynthesis of a floral suppressor .

phyB

phyB plays an inhibitory role in floral initiation. The Arabidopsis phyB mutant flowered earlier than the wild type in both LD and SD conditions, but the early-flowering phenotype of the phyB mutant is more pronounced in SD than in LD conditions. phyB mutations of pea, and sorghum showed early-flowering and decreased photoperiodic sensitivities. More interestingly, in contrast to thephyB mutant of the LD plant pea that flowered early in SD but not in LD, the phyB mutant of the SD plant sorghum flowcred early in LD but not in SD. Therefore, phyB inhibits floral initiation in both LD plants and SD plants, but the phyB inhibition of flowering appears more apparent in the photoperiod that normally suppresses flowering in the respective plant. However, the function of phyB in floral initiation may be more complex than simply as a floral inhibitor. For example, transgenic Arabidopsis plants overexpressing phyB also flowered earlier than the wild type, which could not be easily explained.

phyD

An Arabidopsis phyD mutation was identified as a naturally occurring allele of the wild-type Wassilewskija ecotype, which encoded no functional phyD protein . ThisphyD mutant allele was introgressed into various genetic backgrounds and used to study phyD function. The monogenicphyDmutant plants had no obvious phenotypic abnormality, whereas plants impaired in both the PHYB and thePHYD genes flowered earlier than the phyBmonogenic mutation in both LD and SD conditions (Aukerman et al., 1997; Devlin et al., 1999b). This indicated that, like phyB, phyD inhibits flowering. The triple mutantphyAphyBphyD still retained the ability to respond to the end-of-day far-red light treatment by developing elongated rosette internodes and accelerated flowering—the responses collectively known as the "shade avoidance syndrome" (Smith and Whitelam, 1997; Devlin et al., 1999a; Morelli, 1999). Accelerated flowering under shade, in which there is more far-red light, may allow plants to complete their life cycle before the canopy of other plants becomes too dense. It has been hypothesized that Arabidopsis has at least one other phytochrome associated with shade avoidance responses (Devlin et al., 1999b).

phyE

Based on the hypothesis that Arabidopsis has another phytochrome associated with shade avoidance responses (Devlin et al., 1999b), a genetic screen was carried out to look for mutations that exhibited elongated rosette internodes, resulting in the isolation of thephyE mutation (Devlin et al., 1999a). ThephyEmutant showed no phenotypic alteration unless it was in thephyB mutant background. This indicated the function of phyE is also similar to that of phyB. Among other phenotypes, thephyBphyE double mutant flowered much earlier than thephyB monogenic mutant in SD conditions (Devlin et al., 1999a). In SDconditions, plants containing mutations in both the PHYB and PHYEgenes flowered so early that the end-of-day far-red-light treatment no longer caused further acceleration of flowering (Devlin et al., 1999a). It appears that phyB and phyE normally inhibit flowering in a redundant manner, but their effects can be suppressed by an end-of-day far-red-light treatment. Plants containing mutations in bothPHYB and PHYE genes have reduced suppression of floral initiation, so that they flower early with or without the end-of-day far-red-light treatment. The two Arabidopsis mutants impaired in phytochrome chromophore biosynthesis,hy1 and hy2, flowered earlier than the wild type in both LD and SD conditions . Because the deficiency of phytochrome-chromophore synthesis is likely to indiscriminately affect all phytochromes, it may be expected that the collective outcome of the actions of individual phytochromes would be largely inhibitory with respect to the floral initiation of Arabidopsis. The fact that the majority of Arabidopsis phytochromes play inhibitory roles in flowering appears to be consistent with this view.

cry1

The function of cry1 in flowering seems complicated, although it may have a promotive effect. The hy4 mutant (in the Landsberg erecta ecotype background) was shown to flower late under SD conditions . It was also reported that hy4 mutant alleles in the Columbia ecotype background flowered late in both SD and quasi-LD conditions with either day extensions or night breaks, and that night breaks with blue light had a stronger effect than night breaks with white light or red light (Bagnall et al., 1996). However, in contrast to other photoreceptors, there is a great deal of inconsistency in the flowering time of thecry1 mutant. Some of these inconsistencies may be explained by allele-specific effects. For example, contrary to other cry1mutant alleles,hy4-3 (in the wild-type Wassilewskija ecotype background) andhy4-6 (in the Columbia ecotype background) flowered earlier than the wild type inSD conditions, which was interpreted as being the result of the direct interaction between phyB and the mutant CRY1 protein (Ahmad et al., 1998). However, the inconsistency in flowering time can also be found in reports concerning the identical cry1 mutant allele. The mode of action of cry1 in floral initiation remains unclear.

cry2

cry2 promotes flowering. cry2 mutants are allelic to the photoperiod-hyposensitive late-flowering fha mutant, although the cry2 alleles (in the Columbia ecotype background) had stronger phenotype than the fha alleles. cry2mutant plants flowered late in LD but not in SD conditions, transgenic plants overexpressing cry2 flowered early in SD but not in LD conditions. Therefore, either a mutation or an overexpression of the CRY2 gene resulted in the reduced sensitivity to photoperiods.

Since blue light is known to promote flowering of Arabidopsis, one may expect that the cry2 mutant, which is impaired in a blue light receptor, would show a delayed flowering in blue light. Surprisingly, this was not the case. The cry2mutant flowered at the same time as the wild type in continuous blue light or red light, but the late-flowering phenotype of cry2 in white light could be phenocopied in blue-plus-red light. Therefore, the flowering promotion function of cry2 is dependent on both blue and red light.

Teamwork of Photoreceptors

Why is the late-flowering phenotype of the cry2mutation only revealed in the presence of both blue light and red light? In other words, why does the function of a blue light receptor, cry2, require red light? It was proposed that a cryptochrome may need to be phosphorylated by a phytochrome in red light to become fully active (Ahmad et al., 1998). However, this model may not explain how cry2 regulates flowering time, because genetic studies have demonstrated that phytochromes and cry2 often had opposite effects on floral initiation. Furthermore, the function of cry2 in flowering appeared to require the simultaneous presence of red and blue light, implying that either the red-light-activated cry2 is extremely short-lived, or that red light is not directly required for the biochemical activity of cry2 (Mockler et al., 1999). The latter scenario, as depicted in the double-negative model, predicts that cry2 promotes flowering through its suppression of the phyB-mediated red-light inhibition of floral initiation (Guo et al., 1998; Mockler et al., 1999). Indeed, phyB mutant plants showed a much more pronounced early-flowering phenotype in continuous red light than in continuous white light or continuous red-plus-blue light.

It appears that, similar to its function in hypocotyl elongation, the inhibitory action of phyB in floral initiation is also dependent on red light . Consistent with the hypothesis that phyB inhibits flowering whereas cry2 inhibits phyB action, aphyBcry2 double mutant flowered as early as thephyB mutant in red-plus-blue light. The phyBmutation did not completely suppress the cry2 mutant phenotype in white light (Mockler et al., 1999), which may be explained by the redundant function of phyD and phyE. However, whether phyD- or phyE-mediated inhibition of floral initiation is dependent on red light remains to be investigated.

An antagonistic interaction may also exist between phyB/D/E and phyA—another photoreceptor known to promote flowering.phyAphyB double mutant plants flowered earlier than thephyA mutant, and in certain conditions, the double mutant flowered almost as early as the phyB monogenic mutant. These observations indicate that phyA may also inhibit the function of phyB, and possibly phyD and phyE as well. Because the phyA mutant flowered late in response to a day extension with incandescent light rich in far-red spectra (Johnson et al., 1994), it is tempting to speculate that, analogous to the antagonism between cry2 and phyB and to the far-red-light-dependent phyA function in hypocotyl inhibition, phyA may mediate a far-red-light-dependent inhibition of the phyB function. One test of this hypothesis would be to compare the flowering time of the Arabidopsis wild type with phyAorphyAphyB mutants grown in red-plus-far-red light or in far-red light under conditions allowing plants to flower in the absence of photosynthesis (Araki and Komeda, 1993).

In addition to antagonistic actions, photoreceptors can work in redundant ways to regulate floral initiation. As described previously, phyB, phyD, and phyE inhibit flowering in a redundant manner. Similar redundancy has also been found for cry1 and cry2. Thecry1/cry2 double mutant flowered late in continuous blue light, although neither the monogenic cry1 orcry2 mutant exhibited delayed flowering in such conditions (Mockler et al., 1999). This observation was interpreted to mean that cry2, in addition to its antagonism to phyB, also mediates a blue-light-dependent promotion of floral initiation, but the latter action of cry2 is redundantly carried out by cry1.

Photoreceptors Regulate the Expression of Flowering-Time Genes

How does the action of a photoreceptor affect floral initiation? Given that both phytochrome and cryptochrome can enter the nucleus, there is the possibility that photoreceptors regulate expression of flowering-time genes without invoking second messages for signal transduction. It has been shown that some flowering-time genes are differentially expressed in different photoperiods. For example, the activity of the LEAFY promoter was more quickly up-regulated in LD than in SDconditions . This is significant because floral initiation is determined to a large degree by the level of LEAFY expression (Weigel and Nilsson, 1995). Consistent with the important role of LEAFYon floral initiation and the opposite effect of phyB and cry2 on flowering time, the mutation of the CRY2 orPHYBgenes has been shown to repress or activate LEAFY promoter activity, respectively. Expression of another flowering-time gene, CO, is also dependent on photoperiod, and is expressed at higher levels in LD than in SD conditions (Putterill et al., 1995).

It was reported that CO was expressed at lower levels incry2 mutant plants than the wild type in LD conditions, whereas CO expression was elevated in transgenic plants overexpressing CRY2 in both LD and SD conditions . On the other hand, the expression of CO did not seem to be altered in the phyBmutant . Given the redundant function of phyB/D/E, it will be interesting to see how the expression of CO may be affected in thephyBphyD or phyBphyE double mutant plants. A systematic survey of the expression of more flowering-time genes in various photoreceptor mutations and under different photoperiodic conditions may provide a clearer picture of the role of different photoreceptors in the regulation of expression of the flowering-time genes.

HOW PHOTORECEPTORS REGULATE FLOWERING TIME IN RESPONSE TO PHOTOPERIODS

We have so far conveniently overlooked the question of how photoreceptors regulate flowering time in response to different photoperiods. Apparently, the signal transduction of photoreceptors needs to interact with the circadian clock to regulate flowering time in different daylengths, but the molecular aspects of such interactions remains unclear. It is possible that a photoreceptor regulates the pace and activity of the circadian clock, which in turn regulates floral initiation. Another compelling hypothesis is the external coincidence model, which was initially proposed in the 1930s and later modified to explain why light applied to plants at different times of a dark treatment had different effects on flowering time (Thomas and Vince-Prue, 1997;Carre, 1998). According to this hypothesis, the functions of photoreceptors are 2-fold: first, photoreceptors regulate operation of the circadian clock, and secondly, photoreceptors mediate signal transductions that directly affect floral initiation.

The action of the circadian clock governs, at any given time, the effect of a photoreceptor (or a plant's responsiveness to the light signal) on floral initiation, which often exhibits the photoperiodic response rhythm. Under the appropriate experimental conditions (such as transferring plants grown in a specified photoperiod to continuous darkness and treating them with light at different times), the effect of light on floral initiation may be permitted or denied at certain times of day by the actions of the circadian clock (Carre, 1998). A regulation of the signal transduction of photoreceptors by the circadian clock has been referred to as gating (Millar and Kay, 1996).

Circadian Clock and Regulation of Flowering Time

The circadian clock is an internal oscillator, or it may be more broadly defined as the signaling system that is made up of three functional components—an internal oscillator (or central pacemaker) that generates the circadian oscillation, an input pathway that resets (entrains) the pacemaker according to the environmental cues such as light, and an output pathway that renders oscillations of the pacemaker to overt circadian rhythms (Dunlap, 1999; Somers, 1999). Although the molecular basis of the circadian clock in plant remains unclear, studies in other organisms have established a transcriptional negative feedback loop as the essential component of the central pacemaker . Several Arabidopsis flowering-time genes have been recently isolated and shown to be associated with the function of the circadian clock. Mutations or misexpression of these genes resulted in a decreased photoperiodic response of flowering, which provided the direct evidence for the essential role of the circadian clock in the regulation of photoperiodic flowering. The clock-related genes known to affect flowering time include ELF3, TOC1,CCA1, LHY, and GI.

elf3 was isolated as a photoperiod-hyposensitive early-flowering mutation that flowered early in both LD andSD conditions, and it also exhibited an elongated hypocotyl in red and blue light (Zagotta et al., 1996). Theelf3 mutant lacked circadian rhythms for bothCAB2 promoter activity and leaf movement when assayed under constant light, but the elf3 mutant retained rhythmicity in constant dark, suggesting a possible function of ELF3 in the input pathway (Hicks et al., 1996).

The Arabidopsis toc1 was isolated as a circadian clock mutation that has a short period in every overt rhythm analyzed (Somers et al., 1998b). The effect of daylength on flowering time was diminished in the toc1 mutant in the C24 ecotype background, and was nearly eliminated when the toc1 mutation was introgressed into the Landsberg erecta background. It is particularly interesting that the toc1-1alleles of the C24 ecotype flowered earlier or later than the wild type in SD or LD conditions, respectively (Somers et al., 1998b), whereas almost all other flowering-time mutations affect flowering time in only one direction (either early or late). Isolation of the TOC1 gene may provide more insight into photoperiodic flowering.

CCA1 and LHY both encode MYB-related transcription factors, which when overexpressed cause photoperiod-hyposensitive late flowering, elongated hypocotyls in white light, and disrupted overt circadian rhythms in Arabidopsis (Schaffer et al., 1998; Wang and Tobin, 1998). The expression of CCA1and LHYboth showed circadian rhythms that could be abolished by the overexpression of the respective gene. The Arabidopsiscca1 loss-of-function mutant showed a shortened period length of circadian expression of several genes . CCA1 and LHY may function, in a partially redundant manner, in the regulation of the circadian clock. CCA1 and LHY are regulated by the protein kinase CK2, because CK2 has been shown to interact and phosphorylate both CCA1 and LHY . Overexpression of the CKB3 gene, which encodes a regulatory subunit of CK2, resulted in increased CK2 activity, shortened periods of many clock-related genes, and photoperiod-hyposensitive early flowering .

Another photoperiod-hyposensitive late-flowering mutation in Arabidopsis is gi(Koornneef et al., 1998). TheGI gene was cloned recently and shown to encode a putative membrane protein (Fowler et al., 1999; Park et al., 1999). It is very interesting how a membrane protein like GI may affect the time of floral initiation, which had previously been thought to be determined largely by the regulation of the expression of flowering-time genes. The expressions of GI,CCA1, LHY, andELF3 were found to be dependent on each other (Fowler et al., 1999; Park et al., 1999). The expression of GI exhibited a circadian rhythm with different cycling phases in LD and SD conditions . It is conceivable that the circadian expression of other flowering-time genes, including CCA1, LHY,ELF3, and CO, may also have distinct cycling phases in different photoperiods.

The function of these clock-related genes may directly affect the floral initiation process. Alternatively, these genes may act as the gating factors to regulate the signal transduction of a photoreceptor, as predicted by the external coincidence model. Elucidation of how these proteins affect flowering time will likely significantly enhance our understanding of the photoperiodic flowering.

Photoreceptors and the Entrainment of the Circadian Clock

The circadian clock is entrained by the action of photoreceptors to oscillate with a period of about 24 h. In light, the action of photoreceptors generally accelerates the pace of the clock, resulting in shortened period length comparing to that in dark . It has been shown that mutations of photoreceptor genesPHYA, PHYB, and CRY1 causes the circadian rhythm of CAB2 promoter activity to oscillate at a pace slower (with a longer period length) than that of the wild type under various light conditions (Somers et al., 1998a). This study revealed that in the regulation of the Arabidopsis circadian clock, phyA acts in low intensities of red light and blue light, phyB functions in high-intensity red light, and cry1 acts in both low and high intensities of blue light. The function of phyA in the entrainment of the circadian clock in response to blue light was further demonstrated by showing that the phyA mutant was slower in adapting to a new light/dark condition in low- but not in high-fluence blue light compared with the wild type. Interestingly, thecry2 mutation, despite its reduced sensitivity to photoperiod, did not significantly affect the circadian clock, at least when it was measured for the CAB2 promoter activity (Somers et al., 1998a). This result is consistent with a view that cry2 may not have a major role in mediating light regulation of the circadian clock.

Although phyA, phyB, and cry1 are clearly involved in the regulation of the circadian clock, it is difficult to distinguish whether the abnormality in flowering time observed in the phyA,phyB, and cry1 mutants is the consequence of the malfunction of regulation of the circadian clock, a manifestation of the direct action of the respective photoreceptor on the floral initiation process, or both. It is interesting that mutations in thePHYA, PHYB, and CRY1 genes affected the circadian clock in the similar manner (they all caused longer period length for the circadian expression of the CAB2promoter), yet their effects on flowering time were dissimilar and sometimes opposite (*e.g.* the phyA mutant flowered late but the phyB mutant flowered early). This phenomenon seems to suggest that the observed alterations in flowering time of thephyA, phyB, and cry1 mutants are unlikely to be the direct consequence of a malfunction of the circadian clock. Instead, these photoreceptors may directly affect the floral initiation process, but the signal transduction of photoreceptors may be gated (rather than executed) by the circadian clock, as predicted by the external coincidence model.

The Gated Signal Transduction Paths of Photoreceptors

The function of the circadian clock in regulating flowering time can be demonstrated by the photoperiodic response rhythm or the circadian periodicity of floral induction (or inhibition) in response to light treatment applied at different times of the day (Thomas and Vince-Prue, 1997). For example, Arabidopsis plants grown inSD conditions could be promoted to flower by a 3-h far-red-light treatment applied at various times during a 3-d dark period, and the promotion of flowering by such treatments exhibited circadian rhythms (Carre, 1998). This observation can be explained by an external coincidence model, that the action of a photoreceptor on floral initiation is gated by the circadian clock. It will be interesting to determine whether phyA mediates this far-red-light response. It will also be useful to systematically investigate the photoperiodic response rhythms for other spectra of light in various photoreceptor mutations.

The phyB-regulated floral initiation may represent another example for gated photoreceptor signal transduction. phyB mutations of the SD plant sorghum and the LD plant Arabidopsis (or pea) both caused an early-flowering phenotype. This was somewhat surprising given that the flowering of SD and LD plants responds oppositely to daylength. One interpretation of this observation is that phyB action may suppress floral initiation regardless of photoperiods, but the signal transduction or cell's responsiveness to phyB signaling is gated by the action of the circadian clock, resulting in different daylength responses in the flowering time of different plants.

Finally, the effect of cry2 on floral initiation may be best interpreted by the gating hypothesis. As described previously, the function of cry2 is clearly involved in photoperiodic flowering, because both mutation and overexpression of the CRY2gene result in reduced responsiveness of floral initiation to photoperiods. However, unlike phyA, phyB, and cry1, cry2 is not obviously involved in light entrainment of the circadian clock. Therefore, the effect of cry2 on photoperiodic flowering is more likely to result from its signal transduction (or the plant's response to cry2 signaling) being differentially affected by an output of the circadian clock in different daylength conditions.

CONTROL OF FLOWERING TIME

Plant development is often synchronised to the changing seasons. Flowering, tuberisation and onset of bud dormancy all occur at appropriate times of the year. Changing daylength is one of the environmental signals used by plants to trigger these developmental processes, as was first shown in the 1920s. The ability to monitor and respond to daylength, called photoperiodism, is now known to be widespread throughout the plant and animal Kingdoms. This response is an important character in adaptation of both wild species and crop plants to life at different latitudes. My research group is focused on understanding the molecular mechanisms used by plants to detect daylength and to use this information to trigger flowering. We are also interested in how this process interacts with responses to other environmental signals such as temperature, and how the mechanisms underlying daylength responses are modified in other species to generate responses distinct to those shown by Arabidopsis.

So far, most of our work has focused on the isolation and characterization of genes that regulate flowering of the model species Arabidopsis thaliana. Flowering of Arabidopsis occurs rapidly under long days of 16 hours light, but is dramatically delayed under short days of 10 hours light. We and others have described genetic and molecular interactions between a set of genes that defines a pathway that promotes flowering in response to long days (reviewed in Mouradov et al, 2002). We are now extending our understanding of this long-day pathway by identifying more genes that act within it, seeking to explain the molecular interactions between the gene products at the level of biochemistry and describing in which plant tissues the different components of the pathway act to promote flowering. We are also beginning to investigate whether the long-day pathway occurs in plant species unrelated to Arabidopsis that flower in response to short days, and if so how its activity is modified to create these distinct responses.

CONSTANS - A LIGHT SENSITIVE TIMER

The most extensively studied of the Arabidopsis flowering time genes that confer a daylength response is called CONSTANS (CO). Inactivation of the gene causes late flowering, while its overexpression induces early flowering (Putterill et al, 1995; Simon et al, 1996; Onouchi et al, 2000).

The CO protein is located in the nucleus and is involved in the regulation of the transcription of other genes that regulate flowering time (Robson et al, 2001). For example, CO directly activates the transcription of the flowering time gene FT, which encodes a regulator of phosphorylation (Samach et al, 2000). Although the level of FT transcription is extremely sensitive to CO levels and is induced rapidly in response to initiating CO expression, CO is not predicted to bind directly to the FT promoter but probably acts within a protein complex that includes a DNA binding protein that enables attachment of the complex to the FT promoter.

In wild-type plants, CO activates FT expression under long but not short days, so how does daylength regulate the activity of CO? Progress in answering this question came from the demonstration that CO is expressed only at certain times of day (Suarez-Lopez et al, 2001). These times correspond to the interval when plants are exposed to light in long days, but under short days are in darkness. Therefore, activation of CO protein function by light may explain how its activity is restricted to long days. We obtained further support for this model by studying the environmental conditions under which CO will activate FT transcription. Transgenic plants that express CO mRNA at high level from a viral promoter also express very high levels of FT mRNA, as long as the plants are exposed to light, but when these plants are shifted to darkness, FT transcription declines rapidly. This suggests that exposure to light is required to activate CO protein function.

Our studies on CO explain in broad terms how diurnal rhythms in transcription and post-transcriptional regulation of protein function by light could trigger flowering in response to daylength. We are working intensively to identify the signal transduction pathway that triggers CO activity in response to light, and to understand the effect of exposure to light on CO protein.

GENERATING DIURNAL PATTERNS IN CO MRNA ABUNDANCE

The importance of the diurnal patterns in CO transcription for the control of flowering in response to daylength, led us to study the mechanisms by which these patterns are generated. Other flowering-time genes that act within the daylength pathway are required for these diurnal patterns in CO expression. For example, the gigantea (gi) and lhy-1 mutations both cause late flowering, and these mutations reduce the abundance of the CO mRNA at times when it reaches peak levels in wild- type plants (Suarez-Lopez et al, 2001). That these reductions in CO mRNA are the basis of the late flowering phenotype was further suggested by showing that overexpression of CO in these mutant backgrounds corrected the late flowering phenotype associated with the mutations.

We have studied the functions of the LHY and GI genes in some detail (Schaffer et al, 1997; Fowler et al, 1999). LHY encodes a MYB transcription factor that, along with the related protein CCA1, controls diurnal rhythms in a wide range of Arabidopsis genes, particularly those, like CO, that peak late in the day . When plants are removed from daily cycles of light and dark into continuous light then LHY and CCA1 are required to maintain the rhythms in all circadian clock regulated genes tested. We continue to study the roles of LHY and CCA1 in generating diurnal rhythms, and particularly their function in regulating the expression of genes such as GI and CO that control flowering time.

INTERACTIONS BETWEEN PHOTOPERIOD AND TEMPERATURE

Seasonal responses are often mediated by temperature as well as photoperiod. For example, many plants that grow at high latitude or high altitude will not flower until they have been exposed to low temperatures for several weeks. This requirement for extended exposure to cold is called vernalization and ensures that plants flower in the spring, when optimal conditions for seed set prevail, and not in autumn when seed maturation would not occur prior to the onset of extreme winter conditions. Arabidopsis varieties that require vernalization will flower early if vernalized and very late if they are not vernalized. This late flowering in the absence of vernalization occurs even in long photoperiods, and therefore overrides the promotion of flowering caused by long days.

We have become interested in the mechanism by which the promotion of flowering caused by CO and other long-day pathway genes is suppressed in varieties that require vernalization. Other groups have characterized in detail the genes that are present in strains requiring vernalization, and shown that a MADS box transcription factor called FLC is expressed at high levels in these strains and represses flowering. However, on vernalization FLC expression falls and the plant flowers early. We have shown that FLC suppresses the ability of CO to activate its target genes such as FT (Samach et al, 2000). FLC does not appear to influence CO transcription, and will suppresses the capacity of 35S::CO to activate FT and another target gene, SOC1.

We have analysed the promoters of SOC1 and FT and shown that FLC will bind directly to the SOC1 promoter to suppress its transcription, and that this effect cannot be overcome by high level expression of CO. We are using a combination of genetic and biochemical approaches to study the mechanism by which FLC suppresses CO-mediated activation of genes such as FT and SOC1. We are also interested to determine whether this antagonism between FLC, acting in the vernalization pathway, and CO acting in the photoperiod pathway, is conserved in other plant species.

DIVERSITY IN DAYLENGTH RESPONSES

Different plant species show diverse responses to daylength. For example, although flowering of Arabidopsis and crop plants such as wheat is promoted by exposure to long days, flowering in species such as rice and maize show the reverse response being promoted by short days and inhibited by long days. Does the CO gene also control flowering in these short day plants? A Japanese research group recently demonstrated the importance of CO like genes in short-day plants by showing that differences in the control of flowering of rice varieties could be explained by alterations in the structure of the rice orthologue of the CO gene (Hdl), which in rice promotes flowering in response to short days. We wish to explain how diverse responses to daylength in different plant species are generated. We have shown that CO from the short-day dicotyledonous plant Pharbitis nil will activate FT transcription in Arabidopsis.

We wish to compare the protein complexes in which CO acts in Arabidopsis and Pharbitis and how these are involved in regulating FT transcription as an approach to explaining the basis of the differences underlying long and short day responses.

A ROLE FOR SUMOYLATION IN FLOWERING TIME CONTROL

We have recently become interested in the role of the ubiquitin-like molecule SUMO (small ubiquitin related modifier) in plants, and particularly in the control of flowering. Previously we identified a gamma-ray induced mutation that caused an extreme early flowering phenotype under short days. We cloned the gene by map- based cloning, and showed that it is a plant homologue of ULP1, a cysteine protease that is specific for the ubiquitin-like protein SUMO (also called sentrin). ULP1 both cleaves the carboxy-terminal end of SUMO to generate the mature form, and deconjugates SUMO from target proteins. We have shown that the Arabidopsis protein will cleave SUMO precursor in vitro, and that in vivo the levels of SUMO are severely depleted in the mutant. We are introducing transgenes designed to alter SUMO metabolism in wild-type and mutant plants, with the aim of identifying SUMO-conjugated proteins in Arabidopsis that are involved in the regulation of flowering.

4

Floriculture

Floriculture, or flower farming, is a discipline of horticulture concerned with the cultivation of flowering and ornamental plants for gardens and for floristry, comprising the floral industry. The development, via plant breeding, of new varieties is a major occupation of floriculturists.

Floriculture crops include bedding plants, houseplants, flowering garden and pot plants, cut cultivated greens, and cut flowers. As distinguished from nursery crops, floriculture crops are generally herbaceous. Bedding and garden plants consist of young flowering plants (annuals and perennials) and vegetable plants. They are grown in cell packs (in flats or trays), in pots, or in hanging baskets, usually inside a controlled environment, and sold largely for gardens and landscaping. Pelargonium ("geraniums"), Impatiens ("busy lizzies"), and Petunia are the best-selling bedding plants. The many cultivars of Chrysanthemum are the major perennial garden plant in the United States.

Flowering plants are largely sold in pots for indoor use. The major flowering plants are poinsettias, orchids, florist chrysanthemums, and finished florist azaleas. Foliage plants are also sold in pots and hanging baskets for indoor and patio use, including larger specimens for office, hotel, and restaurant interiors.

Cut flowers are usually sold in bunches or as bouquets with cut foliage. The production of cut flowers is specifically known as the cut flower industry. Farming flowers and foliage employs special aspects of floriculture, such as spacing, training and pruning plants for optimal flower harvest; and post-harvest treatment such as chemical treatments, storage, preservation and packaging. In Australia and the United States some species are harvested from the wild for the cut flower market.

FLORICULTURE IN INDIA

Floriculture is an age old farming activity in India having immense potential for generating gainful self-employment among small and marginal farmers. In the recent years it has emerged as a profitable agri-business in India and worldwide as improved standards of living and growing consciousness among the citizens across the globe to live in environment friendly atmosphere has led to an increase in the demand of floriculture products in the developed as well as in the developing countries worldwide. The production and trade of floriculture has increased consistently over the last 10 years. In India, Floriculture industry comprises flower trade, production of nursery plants and potted plants, seed and bulb production, micro propagation and extraction of essential oils. Though the annual domestic demand for the flowers is growing at a rate of over 25 per cent and international demand at around ₹ 90,000 crore India's share in international market of flowers is negligible. However, India is having a better scope in the future as there is a shift in trend towards tropical flowers and this can be gainfully exploited by country like India with high amount of diversity in indigenous flora.

After liberalization the Government of India identified floriculture as a sunrise industry and accorded it 100 percent export oriented status. The liberalization of industrial and trade policies paved the way for the development of export oriented production of cut flowers. The new seed policy has already made it feasible to import planting material of international varieties. Floriculture products mainly consist of cut flowers, pot plants, cut foliage, seeds bulbs, tubers, rooted cuttings and dried flowers or leaves. The important floricultural crops in the international cut flower trade are rose, carnation, chrysanthemum, gerbera, gladiolus, orchids, anthurium, tulip and lilies.

According to statistics indicated in the Handbook on Horticulture Statistics 2014, the total area under flower crops in 2012-13 was 232.70 thousand hectares. Total area under floriculture in India is second largest in the world and only next to China. Production of flowers was estimated to be 1729.2 MT of loose flowers and 76731.9 million (numbers) of cut flowers in 2012-13. Fresh and Dried cut flowers dominate floriculture exports from India.

Among states, Karnataka is the leader in floriculture with about 29,700 hectares under floriculture cultivation. Other major flower growing states are Tamil Nadu and Andhra Pradesh in the South, West Bengal in the East, Maharashtra in the West and Rajasthan, Delhi and Haryana in the North.

The expert committee set up by Govt. of India for promotion of export oriented floriculture units has identified Bangalore, Pune, New Delhi and Hyderabad as the major areas suitable for such activity especially for cut flowers. Of the four zones identified as potential centers for flower production namely Bangalore, Hyderabad, Pune and New Delhi, the area around Bangalore and Pune have got the advantage of ideal climatic conditions where the temperature ranges between 15 to 30°C. In view of this, the units established in these locations do not require either cooling or heating system. As a result maximum number of units has been established in these locations. There are more than 300 export oriented units in India. APEDA (Agricultural and Processed Food Products Export Development Authority) is the registering authority for such units.

MARKETING

In India Marketing of cut flowers is much unorganized. In most of the Indian cities flowers are brought to wholesale markets, which mostly operate in open yards. From here the flowers are distributed to the local retail outlets which more often than not operate in the open on-road sides, with different flowers arranged in large buckets. In the metropolitan cities, however, there are some good florist show rooms, where flowers are kept under controlled temperature conditions, with considerable attention to value added service.The government is now investing in setting up of auction platforms, as well as organized florist shops with better storage facilities to prolong shelf life. The packaging and transportation of flowers from the farms to the retail markets at present is very unscientific. The flowers, depending on the kind, are packed in gunny bags, bamboo baskets, simple cartons or just wrapped in old newspapers and transported to markets by road, rail or by air. However, the government has provided some assistance for buying refrigerated cargos and built up a large number of export oriented units with excellent facilities of pre-cooling chambers, cold stores and reefer vans.

According to a study titled, 'Indian Floriculture Industry: The Way Ahead' released by the apex industry body ASSOCHAM, India's floriculture industry is growing at a compounded annual growth rate of about 30 per cent, and is likely to cross ₹ 8,000 crore mark by 2015. Currently, the floriculture industry in India is poised at about ₹ 3,700 crore with a share of a meagre 0.61 per cent in the global floriculture industry which is likely to reach 0.89 per cent by 2015.

EXPORT CONSTRAINTS

In spite of an abundant and varied production base, India's export of floricultural product is not encouraging. The low performance is attributed to many constraints like non-availability of air space in major airlines. The Indian floriculture industry is facing with a number of challenges mainly related to trade environment, infrastructure and marketing issues such as high import tariff, low availability of perishable carriers, higher freight rates and inadequate refrigerated and transport facilities.

At the production level the industry is faced with challenges mostly related to availability of basic inputs including quality seeds and planting materials, efficient irrigation system and skilled manpower. In order to overcome these problems, steps must be taken to reduce import duty on planting material and equipment, reduce airfreight to a reasonable level, provide sufficient cargo space in major airlines and to establish model nurseries for supplying genuine planting material. Training centres should be established for training the personnel in floriculture and allied areas. Exporters should plan and monitor effective quality control measures right from production to post harvesting,storage, and transportation.

GOVERNMENT PROGRAMMES AND POLICIES

Department of Agriculture and Cooperation under the Ministry of Agriculture is the nodal organization responsible for development of the floriculture sector. It is responsible for formulation and implementation of national policies and programmes aimed at achieving rapid agricultural growth through optimum utilization of land, water, soil and plant resources of the country. Production of cut flowers for exports is also a thrust area for support. The Agricultural and Processed Food Products Export Development Authority (APEDA), the nodal organization for promotion of agri exports including flowers, has introduced several schemes for promoting floriculture exports from the country. These relate to development of infrastructure, packaging, market development, subsidy on airfreight for export of cut flowers and tissue-cultured plants, database up-gradation etc. The 100 per cent Export Oriented Units are also given benefits like duty free imports of capital goods. Import duties have also been reduced on cut flowers, flower seeds, tissue-cultured plants, etc. Setting up of walk in type cold storage has been allowed at the International airports for storage of export produce.

Initiatives have also been launched for the benefit of exporters by providing cold storage and cargo handling facility for perishable products at various international airports. Direct subsidy up to 50 percent is also available in cold storage units. Besides, subsidy is also provided by APEDA on improved packaging materials to promote their use. To attract entrepreneurship in floriculture sector, NABARD is providing financial assistance to hi-tech units at reasonable interest rates.

Several schemes have been initiated by the Government for promotion and development of the floriculture sector including "Integrated Development of Commercial Floriculture" which aims at improvement in production and productivity of traditional as well as cut flowers through availability of quality planting material, production of off season and quality flowers through protected cultivation, improvement in post harvest handling of flowers and training persons for a scientific floriculture. Many state governments have set up separate departments for promotion of floriculture in their respective states.

Research work on floriculture is being carried out at several research institutions under the Indian Council of Agricultural Research and Council of Scientific and Industrial Research, in the horticulture/floriculture departments of State Agricultural Universities and under the All India Coordinated Floriculture Improvement Project with a network of about twenty (20) centres. The key focus areas are crop improvement, standardization of agro-techniques including improved propagation methods, plant protection and post harvest management. In recent years, however, technologies for protected cultivation and tissue culture for mass propagation have also received attention. A large number of promising varieties of cut flowers have been developed. All these efforts indicate the government's commitment for improving the sector and creating a positive environment for entrepreneurship development in the field.

DEMAND AND SUPPLY

The demand for flowers is seasonal as it is in most countries. The demand for flowers has two components: a steady component and a seasonal component. The factors which influence the demand are to some extent different for traditional and modern flowers.

i. *Traditional Flowers:* The steady demand for traditional flowers comes from the use of flowers for religious purposes, decoration of homes and for making garlands and wreaths. This demand is particularly strong in Kerala, Karnataka, Tamil Nadu, Odisha and West Bengal, as the use of flowers for above mentioned purposes is part of their local culture. The bulk of seasonal demand comes from festivals and marriages. The demand is generally for specific flowers.

ii. *Modern Flowers:* The bulk of the steady demand for modern flowers comes from institutions like hotels, guest houses and marriage gardens. The demand is concentrated in urban areas. With increasing modernization and globalization the demand for modern flowers from the individual consumers is likely to grow enormously as the trend of "say it with flowers" is increasing and the occasions which call for flower giving will continue to present themselves. Although there is an increasing demand for modern flowers from individuals, institutions continue to be the dominant buyers in the market. The price of these flowers also depends on their demand and varies accordingly.

GREEN HOUSE TECHNOLOGY FOR FLOWER PRODUCTION

In present scenario of increasing demand for cut flowers protected cultivation in green houses is the best alternative for using land and other resources more efficiently. In protected environment suitable environmental conditions for optimum plant growth are provided which ultimately provide quality products. Green House is made up of glass or plastic film, which allows the solar radiations to pass through but traps the thermal radiations emitted by plants inside and thereby provide favourable climatic conditions for plant growth. It is also used for controlling temperature, humidity and light intensity inside. On the basis of basic material used, building cost and technology used, green houses can be of three types-

1. *Low-cost greenhouse:* The low-cost green house is made of polythene sheet of 700 gauge supported on bamboos with twines and nails. Its size depends on the purpose of its utilization and availability of space. The temperature within greenhouse increases by 6-10^{0}C more than outside.

2. *Medium-cost greenhouse:* With a slightly higher cost greenhouse can be framed with GI pipe of 15 mm bore. This greenhouse has a covering of UV -stabilized polythene of 800 gauge. The exhaust fans are used for ventilation which are thermostatically controlled. Cooling pad is used for humidifying the air entering the chamber. The greenhouse frame and glazing material have a life span of about 20 years and 2 years respectively.
3. *Hi-tech greenhouse:* In this type of green house the temperature, humidity and light are automatically controlled according to specific plant needs. These are indicated through sensor or signal-receiver. Sensor measures the variables, compare the measurement to a standard value and finally recommend to run the corresponding device. Temperature control system consists of temperature sensor heating/ cooling mechanism and thermostat operated fan. Similarly, relative humidity is sensed through optical tagging devices. Boiler operation, irrigation and misting systems are operated under pressure sensing system.
4. This modern structure is highly expensive, requiring qualified operators, maintenance, care and precautions. However, these provide best conditions for export quality cut flowers and are presently used by large number of export units.

Floriculture has emerged as an important agribusiness, providing employment opportunities and entrepreneurship in both urban and rural areas. National Horticulture Board helps one to establish a flower business. Agricultural and Processed Food Products Export Development Authority helps entrepreneurs with cold storage facilities and freight subsidies. It has been found that Commercial Floriculture has higher potential per unit area than most of the field crops and therefore a lucrative business. During the last decade there has been a thrust on export of cut flowers.

The export surplus has found its way into the local market influencing people in cities to purchase and use flowers in their daily lives. Floriculture thus, offers a great opportunity to farmers in terms of income generation and empowerment. Small and marginal farmers may also use every inch of their land for raising the flower and foliage crops. Floriculture also offers careers in production, marketing, export and research. One can find employment in the floriculture industry as a farm manager, plantation expert, supervisor or project coordinator.

Besides, one can work as consultant or landscape architect with proper training. In addition, floriculture also provides career opportunities in service sector which include such jobs as floral designers, landscape designers, landscape architects and horticultural therapists. Research and teaching are some other avenues of employment in the field.

DEVELOPMENT OF FLORICULTURE :-

Himachal Pradesh is located in North Western part of India between latitude 300 22' 40? N to 330 12' 20? N and longitude 750 45' 55? E to 790 04' 20? N. In Himachal Pradesh, the per capita cultivated land is only 0.12 hectares while per capita irrigated land is a meagre 0.02 hectares. This situation necessitates a cropping pattern that would ensure highest income per unit area/ labour/ investment.

Commercial floriculture perfectly caters to this necessity. The agro-climatic conditions prevailing in the State of Himachal Pradesh offer excellent opportunities for the development of floriculture both to serve the internal off-season market and also for exports, an avenue yet to be tapped. A large variety of floriculture products, *viz.*, cut flowers, bulbs, seeds, live plants, etc. can be produced as economic cash crops.

Although flowers from different agro climatic zones of the State can be made available all through the year for domestic market, export quality flower produce can be ensured only by cultivation under controlled environment conditions of greenhouses.

Table. Commercial Floriculture in Himachal Pradesh Agro Climatic Zones for Floriculture.

Zone description	Elevation range(Meters msl)	Rainfall (cms)	Suitable Flower Crops
Low Hill and Valley Areas near the plains	350 – 900	60 - 100	Gladiolus, Carnation Lilium, Marigold, Chrysanthemum, Rose
Mid Hills(Sub Temperate)	900 – 1500	90 – 100	Carnation, Gladiolus, Lilium, Marigold, Chrysanthemum, Alstroemeria, Rose
High Hills and Valleys in the interiors(Temperate)	1500 – 2750	90 - 100	Gladiolus, Carnation Lilium, Marigold, Chrysanthemum
Cold and Dry Zone(Dry Temperate)	2750 – 3650	24 - 40	Seed/ Corm/ Bulb production

Commercial Floriculture

Gladiolus. Marigold Chrysanthemum

Carnation Rose Lilium

Asiatic Oriental

Potential Floricultural

Alstroemeria Limonium Zantedeschia

Iris Strelitzia Tulips

Gerberas Orchids

Advantages of floriculture :

- The Agro climatic conditions prevailing in the State of Himachal Pradesh offer excellent opportunities for the development of floriculture both to serve the internal off-season market and also exports.
- A large variety of floriculture products, *viz.*, cut flowers, bulbs, seeds, live plants, etc. can be produced.
- The natural agro climatic conditions offer ideal production environment for flowers and the planting material *i.e.*, expensive heating and cooling systems in the greenhouses are not required
- Power required for running the greenhouses is charged at domestic rates in the State.

SERVICES PROVIDED BY THE DEPARTMENT OF HORTICULTURE

INFRASTRUCTURAL SUPPORT

Floriculture Nurseries:

The Department of Horticulture has established seven Floriculture Nurseries in various Districts, *viz.*, Navbahar and Chhrabra in Shimla District, Mahog Bag and Parwanoo in Solan District, Bajaura in Kullu District and Dharamshala and Bhatoon in Kangra District.

Model Floriculture Centre:

The "Model Floriculture Centre" has been established at Mahog Bag (Chail), District Solan and a Tissue Culture Laboratory is being set up for the propagation of planting material of commercially important floriculture crops. The present infrastructure at the "Model Floriculture Centre" consists of 1706.5 sq. m of Greenhouse area, one Handling Unit for post harvest handling of flowers and 3 Nos. of Cool Chambers for forcing and storage of planting material. The building, which shall house the Tissue Culture Laboratory, Training Hall and other infrastructure of the Centre, has been constructed at an estimated cost of ₹ . 94.22 lakhs and taken over by the Department of Horticulture in July 2004.

Post-harvest Infrastructure

Collection, Grading and Packing House and cool chamber facilities have been established by the District Rural Development Agency for post-harvest management of floriculture produce in the districts of Bilaspur, Mandi and Kangra.

Research and Development:

The following organizations provide the necessary R and D support in the field of floriculture: -1. Dr. Y.S. Parmar University of Horticulture and Forestry, Solan. This University has a separate Department of Floriculture and Landscaping as its head quarters at Nauni. The location specific research work is being carried out at the regional Research stations of the university located in various Agro climatic Zones of the State.2. Institute of Himalayan Bio-resource Technology, Palampur, District Kangra3. ICAR Research Station at Katrain District Kullu H.P.4. National Bureau of Plant Genetic Resources, Phagli, Shimla, H.P.

Technical Assistance:

- Training in Floriculture
- Organization of Study Tours
- Advisory Service:Free technical advice is made available to the entrepreneurs and practicing floriculturists in pre and post-harvest technologies of floriculture crops.
- Literature for Floriculture: Literature handouts containing technical information pertaining to cultivation of floriculture crops are supplied free of cost.

Organization of Flower Shows:

The Department provides assistance for the organization of flower shows to create awareness on the usage of floriculture produce – both indoors and outdoors.

- Formation of Flower Growers Co-operative Societies:The Department provides assistance to the flower growers for the formation of flower growers co-operative Societies.

Assistance from other Organizations:

The Department assists the flower grower co-operatives and NGOs to obtain assistance for the establishment of post-harvest management facilities available from organizations like National Horticulture Board, APEDA and NABARD.

Research and Development Support

- Dr. Y.S. Parmar University of Horticulture and Forestry, Solan. This University has a separate Department of Floriculture and Landscaping at its head quarters at Nauni. The location specific research work is being carried out at the Regional Research Stations of the University located in various Agro climatic Zones of the State.
- Institute of Himalayan Bio-resource Technology, Palampur, District Kangra.
- ICAR Research Station at Katrain District Kullu H.P.
- National Bureau of Plant Genetic Resources, Phagli, Shimla, H.P.

Steps Initiated to Promote Floriculture :-

- Creating more public awareness regarding use of floriculture produce through media and other agencies as well as more exposure of floriculture products during consumer exhibitions.
- Retailing of flower produce through super markets in addition to Florist shops to encourage flower consumption especially in metropolitan cities.
- Organizing post harvest infrastructure for marketing needs at the domestic terminal markets, particularly the Delhi market.
- Promotion of interaction between growers and scientific Institutions for effective lab to land technology transfer.

Carnation Production

Protected cultivation of Rose

Chrysanthemum

marigold

FLORICULTURE ACTIVITY: AN OVERVIEW

Floriculture, an important branch of horticulture involves the cultivation of flowers and it includes ornamental gardening and landscaping. Gardening for aesthetic purposes has also been a part of the Indian community with special emphasis on cut flowers. This is an emerging economic field and has been in the forefront in the recent years. Many industrialists and business houses too have expanded their business in the line. The high-tech floriculturists grow flowers under controlled conditions mainly to cater to the demands of the fast expanding international market. Floricultural products constitute a small but important segment of the international trade. These are high value products that are used for their beauty and elegance and they reap very high economic returns when compared to other agricultural and horticultural products.

The floriculture industry in the world showed a dramatic increase in growth during the 1970's and 1980's. The increase in floral products is attributed to standard of living of the people. The major cut flowers that are in demand in the international market are roses, chrysanthemum, carnations, tulips, gerberas, alstroemerias etc. Many new types of flowers like anthuriums, aster, asclepias, spray carnation, gypsophila, liatris, limonium, heliconia etc. that are entering the flora trade.

Over the past decade, flower and pot plant business in the world has increased to US $ 40 billion. The annual rate of growth in the floriculture industry is about 15 percent. The floricultural products include cut flowers, which contribute about60 percent of the global trade, flowering and green potted plants and bedding plants from a small segment of the floricultural crop production worldwide. India's share in this global floriculture market is around 0.75 per cent. There are clear signs that India would soon play an important role in the world trade of flowers. With the liberalization of economic policies and identification of floriculture as one of the thrust areas by the Government opportunities offered by the large global market. In view of the potential for exports, a large number of export oriented units have been approved by the Government of India and many projects have been commissioned in recent years. Domestic market can absorb the non-exportable grades.

To reduce the capital cost and cost of production, it is desirable to establish export oriented units so that concessions on import of capital goods and other consumables can be availed of. J and K State, which has the agro-climate conditions suitable for various varieties of flowers, can be a good source of production and supply of the quality cut flowers to the domestic and international market. State of Jammu and Kashmir have ideal climatic conditions for floriculture. The climate of the three regions of the State *viz.* Jammu, Kashmir and Leh are Sub – tropical, temperate and cold – arid. In J and K State this activity is in vogue mainly in the form of beautification of existing gardens / parks in the State. Department of Garden, Parks and Floriculture develop and maintain parks and gardens at suitable locations to upgrade the aesthetic value. The department is contributing significantly by adding new varieties of flowers and beautification plants in existing gardens and parks. Some potential areas are being developed as demonstration units by the department. State Government proposes to invest an amount of ₹ .

Crores during 9th Five-year plan for the development of this sector. The State intends to propagate floriculture as an important economic activity and hence thrust on its commercial exploitation.

The Department of Garden, Parks and Floriculture is providing technical know-how support for plantation and proper maintenance of plants and also by way of setting up demonstrative parks therein growing greenhouse seedlings, and helps in commercializing this activity through:

Demonstrative units

Training Facilities

Supply of other inputs at Subsidized rate and Market Support The demonstrative units are set up to establish the technical feasibility of the proposal. Training Facilities include providing know how about setting up of green houses and maintaining them, proper growth of cut flowers, handling equipment, post harvest treatment of cut-flowers and the packaging practices. The department also provides subsidy on purchase of bulb for gladilieus and Tulips (50 per cent of its cost) for the maximum purchase of 2000 bulbs by individual private grower. Department also guides and assist the entrepreneurs in marketing of the cut flowers. Department has linkage with departments and institutions engages in marketing of floriculture products at National and International level including Agriculture product Export Development Agency (APEDA), New Delhi. J and K State Industrial Development Corporation extends the promotional support as well as commercial banks / NABARD provides financial assistance for floriculture project.

25 number of floriculture growers has set up their units under the technical guideline and supervision by the experts of the Department. The marketing of cut-flowers could be tied up in European and Gulf Countries where demand of Lily, Rose, Tata Rose, Gladilieus cut flower and other varieties is quite high, The State of Jammu and Kashmir has suitable climate conditions to grow the flowers on commercial lines for National and International markets. Under green house polyhouse conditions, the flowers can be grown through out the year.

The main varieties, which are grown in the State, include Gladilieus, Roses, Tata Roses, Lilly, Tulip, Hyceanthus, Carnation etc. Details pertaining to growing different varieties and suitable locations are as under:

GLADILIEUS

This variety could be grown for cut – flower purposes in both Jammu and Kashmir regions. It is grown ideally within 90 days from the date of seeding of tuber. Ideal locations for cultivation are Jammu, Kud, Chenani, Poonch and Rajouri Districts in Jammu and Srinagar, Budhgam and Anantnag in Kashmir region. In Leh region such flowers can be grown under greenhouse cultivation. The total production of this variety of cut – flowers in the State is estimated at about one lakh numbers.

ROSE

Rose is grown in the month of November starts production from the month of April. Such plant can be grown in greenhouses under controlled temperature between 20 Deg. C. to 30 Deg. C. and also in open field. The output of Roses in greenhouses is around double compared to open field cultivation. The rose flower output is very less in comparison to its existing demand.

LILLY

The demand for this variety of flowers emanates from as well as Delhi and Mumbai markets during Christmas. The bud is ready within 75 days. The suitable locations are Panchari, Sudhmahadev (during winter), Poonch and

Rajouri (during summer) in Jammu region and Srinagar, Budhgam, Pulwama in Kashmir region. The production could be carried out through out the year in Jammu region for 3 to 4 months in Kashmir. The demand is more than 1.5 lakh to 2 lakhs tubes and presently only 10 thousand bulbs is grown in the State.

TULIPS

This variety of flower is grown in Kashmir Region of the State. It is sown in the month of July or August and its spike is ready in around 70 to 75 days time. The spike can be kept for 8 – 9 days and mostly sold during the Christmas. About 40 to 50 thousand tubes is grown and some no. Of spikes are exported to Delhi Market from Kashmir region. However, demand is very high compared to its supply.

HYSENTHESIS

Its production in the State is merged. The potential for the growth of this variety is available in the State. Kashmir, Leh and hilly regions in Jammu, Doda, Poonch district of Jammu region. More than 3 to 3.5 lakhs tubes could be grown in the three locations of State.

MARIGOLD

There is an increasing demand of this variety of flowers by the visiting pilgrims (more than 4.2 millions) at Vaishno Devi Ji. Its total production is estimated at 600 to 700 quintals and unfulfilled demand of the State is of the order of 800 quintals. This variety of flowers are exported during the month of October to November from the State and imported to the State from Delhi etc. in July and August.

IMPACT OF GLOBAL CLIMATE CHANGE ON FLORICULTURE IN INDIA

Climate change is one of the most important global environmental challenges in the history of mankind. It is mainly caused by increasing concentration of Green House Gases (GHGs) in the atmosphere. In 1980s, scientific evidences linking GHGs emission due to human activities causing global climate change, started to concern everybody. Subsequently, United Nations General Assembly in 1992 formed Intergovernmental Negotiating Committee for Framework Convention on Climate Change (UNFCCC) which finally adopted the framework for addressing climate change concerns. The Intergovernmental Panel on Climate Change (IPCC) has been publishing periodic assessment reports on atmospheric carbon concentration and its likely impact on the environment. The IPCC in its 4th Assessment Report states that emission of global GHGs has increased since pre-industrial times, with an increase of 70 per cent between 1970 and 2004. The big challenge before the international community is to limit the emission of green house gases by 2050 and measurably by 2020. Climate of the planet earth is always in a state of change as a natural process influenced by both natural variability and induced environmental changes due to anthropogenic reasons. Natural causes include continental drift, volcanoes, earth's tilt, and ocean current while human causes are GHGs, agricultural practices, energy sources, waste disposal, depleting forest cover, etc. However, the reason for worry is that climate change is taking place at a much faster rate than expected by the human interference. The consequences of such rapid change are - global warming, change of seasonal pattern, excessive rain, melting of ice cap, flood, rising sea level, drought, etc. leading to extremity of all kinds. The implications will be wide spread but specially on the food production (agriculture / horticulture), forest ecosystem, health, energy, etc. Vulnerability, rarity and rapid extinction of plant species will be among other consequences.

Plants are key components of the ecosystem and are greatly influenced by climatic and geographical factors. Therefore, climate change has a direct impact on agriculture and horticulture as the basic factors for crop production are being influenced. Overall, a low production of horticultural crops is feared due to the climate change. Assuming a global temperature rise of 4.4°C by 2080 over the cultivated areas, India's agricultural output is projected to fall by 30-40 per cent which would be quite alarming unless proper remedial measures are taken. Further, occurrence of new diseases, pests together with severity of the existing ones is also foreseen. Some of the well established commercial varieties of fruits, vegetables and flowers will perform poorly in an unpredictable manner.

FLORICULTURAL SCENARIO

India is becoming a strong centre of commercial floriculture in the international market. During the last 5-7 years, there was a great surge in the floricultural activity in the production of flowers (cut and loose), ornamental plants (potted and cut-greens) and dry flowers (value added products), besides marketing. The horticultural sector contributed around 28 per cent of the GDP annually from 13.08 per cent of the area and 37 per cent of the total exports of agricultural commodities (2004-05). Albeit, India's present contribution in the global floricultural export market is negligible (about 0.4 per cent) as compared to Netherlands (58 per cent), Columbia (14 per cent), Ecuador (7 per cent), Kenya (5 per cent), Israel (2 per cent), Italy (2 per cent), Spain (2 per cent) and others 10 per cent, it is not far when India will come up as a major grower/ exporter by virtue of well planned policies formulated by the Government of India backed with foreign technologies for green house production.

IMPACT OF CLIMATE CHANGE ON FLORICULTURE

The impact of climate change on flowering plants and crops will be more pronounced. Melting of ice cap in the Himalayan regions will reduce chilling required for the flowering of many of the ornamental plants like Rhododendron, Orchid, Tulipa, Alstromerea, Magnolia, Saussurea, Impatiens, Narcissus etc. Some of them will fail to bloom or flower with less abundance while others will be threatened. Indigenous species in the natural habitat will be under threat for not getting favourable agro-climatic conditions for their proliferation. Western Ghats and surrounding regions may be deprived of normal precipitation due to abnormal monsoon. Plant species requiring high humidity and water may find them under difficult conditions for survival. Plains of India will also have similar kind of problems and will be affected either by drought or excessive rains, floods and seasonal variations.

Commercial production of flowers particularly grown under open field conditions will be severely affected leading to poor flowering, improper floral development and colour besides reduction in flower size and short blooming period.

FUTURE STRATEGIES

In view of these problems, horticulturists will have to play a significant role in the climate change scenario and proper strategies have to be envisaged for saving horticulture/floriculture from future turmoil. The most effective way to address climate change is to adopt a sustainable development pathway, besides using renewable energy, forest and water conservation, reforestation etc. Awareness and educational programmes for the growers, modification of present horticultural practices and greater use of green house technology are some of the solutions to minimize the effect of climate change. Hi-tech horticulture is to be adopted in an intensive way. It is necessary that selection of plant species/cultivars is to be considered keeping in view the effects of climate change. The performance of different seasonals may not be satisfactory due to shorter and warmer winter. Judicious water utilization in the form of drip, mist and sprinkler will be a key factor to deal with the drought conditions. Development of new cultivars of floricultural crops tolerant to high temperature, resistant to pests and diseases, short duration and producing good yield under stress conditions, will be the main strategies to meet this challenge.

5

Cut Flower Production in Asia

CUT FLOWER PRODUCTION IN INDIA

India has a long tradition of floriculture. References to flowers and gardens are found in ancient Sanskrit classics like the Rig Veda (C 3000-2000 BC), Ramayana (C 1200-1300 BC), Mahabharata (prior to 4th Century BC), Shudraka (100 BC), Ashvagodha (C 100 AD), Kalidasa (C 400 AD) and Sarangdhara (C 1200 AD). The social and economic aspects of flower growing were, however, recognized much later.

The offering and exchange of flowers on all social occasions, in places of worship and their use for adornment of hair by women and for home decoration have become an integral part of human living.

With changing life styles and increased urban affluence, floriculture has assumed a definite commercial status in recent times and during the past 2-3 decades particularly.

Appreciation of the potential of commercial floriculture has resulted in the blossoming of this field into a viable agri-business option. Availability of natural resources like diverse agro-climatic conditions permit production of a wide range of temperate and tropical flowers, almost all through the year in some part of the country or other. Improved communication facilities have increased their availability in every part of the country. The commercial activity of production and marketing of floriculture products is also a source of gainful and quality employment to scores of people.

PRESENT SITUATION OF CUT FLOWER PRODUCTION

Inspite of the long and close association with floriculture, the records of commercial activity in the field are very few. The information on the area under floriculture and the production generated is highly inadequate. As commercial floriculture is an activity which has assumed importance only in recent times, there are not many large farms engaged in organised floriculture. In most part of the country flower growing is carried out on small holdings, mainly as a part of the regular agriculture systems.

Production Areas

The estimated area under flower growing in the country is about 65,000 hectares. The major flower growing states are Karnataka, Tamil Nadu and Andhra Pradesh in the South, West Bengal in the East, Maharashtra in the West and Rajasthan, Delhi and Haryana in the North. It must, however, be mentioned that it is extremely difficult to compute the statistics of area in view of the very small sizes of holdings, which very often go unreported. This perhaps would be the reason for unrealistically small areas reported for floriculturally active states like Maharashtra, Uttar Pradesh and Madhya Pradesh.

More than two thirds of this large area is devoted for production of traditional flowers, which are marketed loose *e.g.* marigold, jasmine, chrysanthemum, aster, crossandra, tuberose etc. The area under cut flower crops (with stems) used for bouquets, arrangements etc. has grown in recent years, with growing affluence and people's interest in using flowers as gifts. The major flowers in this category are rose, gladiolus, tuberose, carnation, orchids and more recently liliums, gerbera, chrysanthemum, gypsophila etc.

Table. Area Under Flower Production in India.

State	Area (ha.)
Karnataka	19,161
Tamil Nadu	14,194
West Bengal	12,285
Andhra Pradesh	5,933
Maharashtra	3,356
Rajasthan	1,985
Delhi	1,878
Haryana	1,540
Madhya Pradesh	1,270
Uttar Pradesh	1,000
Others	2,166
Total	64,768

The production of flowers is estimated to be nearly 300,000 metric tonnes of loose flowers and over 500 million cut flowers with stem. In the case of production also, the estimates could be at variance from the actual figures as some of the flowers like rose, chrysanthemum, and tuberose are used both as loose flowers and with stem. It may be mentioned that almost all of the area reported here is under open field cultivation of flowers. Protected cultivation of flowers has been taken up only in recent years for production of cut flowers for exports. The estimated area in production is about 200 hectares, which is likely to increase to over 500 hectares by the year 2000.

Recognising the potential for low cost production for export, in view of cheap land, labour and other resources, several export oriented units are being set up in the country. These projects, located in clusters around Pune (Maharashtra) in the West, Bangalore (Karnataka) and Hyderabad (Andhra

Pradesh) in the South, and Delhi in the North, are coming up in technical collaboration with expertise mainly from Holland and Israel. More than 90 percent of these units are for rose production, on an average size of 3-hectare farm, while some projects for orchid, anthurium, gladiolus and carnation are also being set up. Nearly one third of over 200 proposed projects, have already commenced production and export.

Major Cut Flower Crops

Rose is the principal cut flower grown all over the country, even though in terms of total area, it may not be so. The larger percentage of the area in many states is used for growing scented rose, usually local varieties akin to the Gruss en Tepelitz, the old favourite to be sold as loose flowers. These are used for offerings at places of worship, for the extraction of essential oils and also used in garlands. For cut flower use, the old rose varieties like Queen Elizabeth, Super Star, Montezuma, Papa Meilland, Christian Dior, Eiffel Tower, Kiss of Fire, Golden Giant, Garde Henkel, First Prize etc. are still popular. In recent times, with production for export gaining ground in the country, the latest varieties like First Red, Grand Gala, Konfitti, Ravel, Tineke, Sacha, Prophyta, Pareo, Noblesse. Virsilia, Vivaldi etc. are also being grown commercially.

Gladiolus is the next most important cut flower crop in the country. Earlier it was considered a crop for temperate regions and its growing was restricted to the hilly areas, particularly in the north eastern region, which still continues to supply the planting material to most parts of the country. However, with improved agronomic techniques and better management, the northern plains of Delhi, Haryana, Punjab, Uttar Pradesh, as well as Maharashtra and Karnataka have emerged as the major areas for production of gladiolus.

Tuberose, a very popular cut flower crop in India is grown mainly in the eastern part of the country *i.e.* West Bengal, and also in northern plains and parts of south. Both single and double flower varieties are equally popular. Tuberose flowers are also sold loose in some areas for preparing garlands and wreaths.

The other main cut flower item is orchid. Its production is restricted mainly in the north-eastern hill regions, besides parts of the southern states of Kerala and Karnataka. The main species grown are Dendrobiums, Vanda, Paphiopedilums, Oncidiums, Phalaenopsis and Cymbidiums.

Among the traditional crops grown for loose flowers, the largest area is under marigold, grown all over the country. In most parts of the country only local varieties are grown for generations. African marigolds occupy more area as compared to the small flowered French types. Jasmine flowers in view of its scent are also very popular as loose flowers and for use in garlands and Veni (ornament for decoration of hair by women). The major areas under this crop are in Tamil Nadu, Karnataka in South and West Bengal in East. The varieties are mainly improved clones of Jasminum grandiflorum, J. auriculatum and J. sambac. The chrysanthemum, particularly the white varieties are much in demand as loose flowers during the autumn period of October-December when other flowers like jasmine, tuberose are not available for use in garlands etc. Among other traditional flowers grown in large areas are crossandra in southern states of Tamil Nadu, Karnataka and Andhra Pradesh and aster in Maharashtra.

Research Support

Research work on floriculture is being carried out at several research institutions under the Indian Council of Agricultural Research and Council of Scientific and Industrial Research, in the horticulture/floriculture departments of State Agricultural Universities and under the All India Coordinated Floriculture Improvement Project with a network of about twenty (20) centres. The crops which have received larger attention include rose, gladiolus, chrysanthemum, orchid, jasmine, tuberose, aster, marigold etc. The thrust till recently had been on crop improvement, standardization of agro-techniques including improved propagation methods, plant protection and post harvest management. In view of the fact that most of the cut flower production is being done under open field conditions, the research efforts generally relate to open cultivation. In recent years, however, technologies for protected cultivation and tissue culture for mass propagation have also received attention. A large number of varieties suitable for cut flower use, as well as garden display have been developed. Production technology, particularly the agronomic requirements and control methods for important diseases and insect pests have also been developed. Contribution by the private sector in research activities in floriculture is negligible.

PLANTING MATERIAL

The requirement of planting material to cater to the large area under flower crops, is largely met from domestic production. Since efforts to set up large commercial farms generally suffered due to lack of quality planting material in sufficient quantities, this aspect has received greater attention in recent years in the breeding centres, which are producing sufficient quantity of planting material. Most of the nurseries propagating planting material are in the private sector. In the absence of any mechanism to register nurseries, it is very difficult to ascertain their exact number, but at a very conservative estimate there are more than 100,000 nurseries, spread out all over the country, producing seeds and other planting materials for flower growers. The states with larger numbers of nurseries include Maharashtra, West Bengal, Karnataka and Tamil Nadu. Most of the nurseries are small, with little or no improved facilities like mist propagation unit, green houses/net houses etc. For meeting the demand of flower seeds, several large seed companies have production units in Punjab,

Himachal Pradesh and Jammu and Kashmir in the North, Karnataka in the South and West Bengal in the East. A few of the leading multinational seed companies have tied up with local seed companies or producers for custom production of seeds of their varieties. In the case of bulbous plants, most of the planting material is produced in the north eastern hilly regions of West Bengal (Kalimpong) and Sikkim, though for some crops, it is also produced in hilly regions of northern India. The introduction of a revised seed policy by the government of India in 1989 has enabled unrestricted introduction of many new and superior varieties into the country, increasing the variety in the floral basket.

Tissue culture has, in recent years, been recognized as an important tool in agriculture development. With its diverse climatic zones and qualified manpower, India is well placed to exploit the benefit of tissue culture based applications to floriculture crops. Most popular application of tissue culture has been micropropagation using in vitrotechnique for mass multiplication of planting material. Tissue culture plants of ornamentals have found ready acceptance by the commercial growers and their production increased significantly from 130 million plants in 1985-86 to 680 million in 1994-95. At present 30 commercial tissue culture units with annual capacities of 0.5 to 15 million plants each are in operation, resulting in total capacity of about 110 million plants. While most of it is exported, a small percentage of cut flower crops like carnation and gerbera are finding good market within the country.

MARKETING

Marketing of cut flowers in India is very unorganised at present. In most metropolitan cities, with large market potential, flowers are brought to wholesale markets, which mostly operate in open yards. A few large flower merchants generally buy most of the produce and distribute them to local retail outlets after significant mark up. The retail florist shops also usually operate in the open on-road sides, with different flowers arranged in large buckets. In the metros, however, there are some good florist show rooms, where flowers are kept in controlled temperature conditions, with considerable attention to value added service. The government is now investing in setting up of auction platforms, as well as organized florist shops with better storage facilities to prolong shelf life.

The packaging and transportation of flowers from the production centres to the wholesale markets at present is very unscientific. The flowers, depending on the kind, are packed in old gunny bags, bamboo baskets, simple cartons or just wrapped in old newspapers and transported to markets by road, rail or by air. The mode of transportation depends on the distance to the markets and the volume. Mostly, flowers are harvested in the evening time and transported to nearby cities by overnight trains or buses. In recent years, the government has provided some assistance for buying refrigerated carriage vans. A large number of export oriented units have built up excellent facilities of pre-cooling chambers, cold stores and reefer vans and their produce coming for domestic market sales are thus of very good quality and have longer vase life and command higher price. The government programmes for floriculture development include creating common facilities of cool chain in large production areas to be shared on cooperative basis. Formation of growers' cooperatives/ associations are being encouraged.

In view of the unorganized set up, it is difficult to estimate the size of flower trade, both in terms of volume and value. A study conducted in 1989 estimated the trade to be worth ₹ . 2050 million. It is in the period of the last five years or so that this business has really boomed in India, which is reflected in the number of new florist outlets in all cities and increase in the public's purchase of flowers as gifts. This would put the current trade at several times the earlier estimate. A recent study of Delhi market alone put the value of flowers traded on wholesale as ₹ . 500 million.

The loose flowers (traditional crops like marigold, jasmine etc.) are usually traded by weight. The average price of different flowers in major markets varies considerably depending on the period of availability.

Table. Average Market Price for Major Flower Crops.

Flowers	Unit	Price (US$1 = ₹ .40) Rs./kg or doz or each stem
Marigold	kg.	3-60
Jasmine	kg.	15-150
Crossandra	kg.	20-120
Chrysanthemum	kg.	5-25
Tuberose	kg.	5-30
Rose	kg.	6-60
Gladiolus	doz.	20-75
Carnation	doz.	30-75
Gerbera	doz.	36-75
Orchids	each stem	10-45
Liliums	each stem	10-45
Anthuriums	each stem	15-45

The net returns to the growers depend on the packaging and transportation costs. The cut flowers with stem have a limited overall market in terms of volume. The share of cut flowers has almost doubled from 30 to 60 per cent in the last decade.

The value of cut flower export from India has increased twenty five fold during the last five years. With more export oriented units coming into operation, exports are likely to grow further in the coming years. The major share of the export trade is for roses, in addition to orchids, gladiolus etc. The major markets are Europe (Holland, Germany and U.K.) and Japan. The exports of roses to Japan, have really picked up in the three years from ₹ . 360 million in 1993-94 to ₹ . 6090 million in 1995-96. As per the estimates for 1996-97, India has been the largest supplier of roses to Japan (volume wise).

Table. Export of Floriculture Products from India (Rupees in Million).

Item	1991-92	1992-93	1993-94	1994-95	1995-96
Cut Flower Fresh	3.99	10.90	9.96	29.98	100.33
Dried Flowers	64.35	74.31	109.21	195.78	364.56
Live Plants	40.15	30.56	30.52	60.43	81.48
Dried Plants	23.99	25.45	23.72	10.35	35.83
Bulbs, Tubers etc.	12.06	7.83	14.95	11.83	19.21
Total	144.54	149.05	188.36	308.37	601.41

POTENTIAL FOR CUT FLOWER PRODUCTION DEVELOPMENT

The availability of natural resources like favourable and diverse climatic conditions permit production and availability of a large variety of flower crops round the year. Cheap labour leads to reduction in production costs, increasing access of the consumer to good quality flowers at affordable prices, besides increasing our competitiveness in the export markets. Being a new concept in the agri-business, it took some time for scientific commercial flower production to take roots, but with the appreciation of its potential as an economically viable diversification option, its growth is slowly stabilising. The government also has, during the last few years, recognized floriculture as an important segment for developmental initiatives.

Model Floriculture Centres being set up in 11 major production zones, to serve as focal units for development in the region, have a mandate of making available quality planting material, new/improved production technologies and also to provide training in production and post harvest management. There are also special government programmes for area expansion in floriculture with state assistance. The National Horticulture Board, a major developmental agency for horticulture, also makes available finances as soft loan for setting up integrated projects for production and marketing. As mentioned earlier, the government is investing in improving the infrastructure for marketing in the domestic sector.

Production of cut flowers for exports is also a thrust area for support. The Agricultural and Processed Food Products Export Development Authority (APEDA), the nodal organization for promotion of agri-exports including flowers, has introduced several schemes for promoting floriculture exports from the country. These relate to development of infrastructure, packaging, market development, air freight subsidy etc. The 100 per cent Export Oriented Units are also given benefits like duty free imports of capital goods.

All these efforts indicate the government's commitment for improving the sector and creating a positive environment for entrepreneurship development in the field.

CONSTRAINTS IN CUT FLOWER PRODUCTION DEVELOPMENT

Being a new concept, the requirements of scientific and commercial floriculture is not properly understood in the country. The developmental initiatives of the government have to keep in mind the low knowledge base, small land holdings, unorganized marketing and poor infrastructural support.

While long experience of flower growing in the open field conditions enable sufficient flower production for domestic markets, the quality of the produce, in view of its exposure to various kinds of biotic and abiotic stresses, is not suitable for the ever growing export market. The production technology for flowers under protected environment of green houses needs to be standardized. There is hardly any post harvest management of flowers for the domestic market. Availability of surplus flowers from exports for sale in the domestic market, has increased the appreciation of quality produce and the demand for good quality flowers is increasing. With the introduction of new varieties of crops in the country, facilities for generating their planting material for large scale production need strengthening. Special attention needs to be paid to strengthen the marketing infrastructure like organised marketing yards, auction platforms, controlled condition storage chambers etc.

Greater research efforts are also needed for integrated pest management, development of location specific package of practices for traditional flowers, value addition to traditional flowers etc. The initial cost and availability of finance is a critical matter in the development of large commercial projects requiring heavy investments. More options for developmental finance, such as the soft loan scheme of the National Horticulture Board need to be identified. In the initial years of commercial floriculture development, the governmental support in terms of subsidies etc. needs special attention.

The potential for growth of export market is always linked to the strength of domestic market - its capacity to absorb surplus and over production, and quality consciousness of consumers. Though we have a large domestic market, the marketing system and facilities need to be modernized.

The production for exports at present has suffered due to a few constraints. While our growers have been successful in producing world class quality at low cost, high air freight rates, low cargo capacity available, imposition of import duties, inadequate export infrastructure etc. have reduced their competitiveness. There is also a shortage of trained manpower to handle commercial floriculture activity. The demands of the growing export oriented industry would require adequate attention to be paid for human resource development, particularly at the supervisory level.

CONCLUSIONS

India has a long floriculture history and flower growing is an age old enterprise. What it has lacked is its commercialization. The growing demands of flowers in the domestic as well as the export market will require a concerted effort on the part of the government as well as the private entrepreneurs to develop floriculture on scientific lines. Paying attention to the input needs, better resource management and making various policies entrepreneur friendly would lead to a balanced growth of the industry.

Table. Cut Flower Exports from India.

Country	1993-94	1994-95	1995-96
Japan	322.50	8255.58	35932.56
Netherlands	1004.61	9102.49	24799.90
U.S.A.	1175.38	2495.21	17652.50
Germany	957.61	2538.63	9256.00
U.K.	1420.93	1113.78	3345.86
U.A.E.	2120.40	3388.19	2459.29
Italy	210.28	164.96	2200.49
Hongkong	730.02	903.15	1504.02
Singapore	78.45	437.37	1190.54
Nepal	11.86	36.09	292.64
Kuwait	24.22	5.00	274.13
Saudi Arabia	413.98	169.52	272.77
Switzerland	136.96	258.20	242.83
Hungary	-	286.69	181.82
Thailand	-	86.49	177.82
Australia	-	-	132.89
Russia	368.46	20.96	119.85
Others	988.68	719.39	9293.82
Total	9964.34	29981.60	109329.73

CUT FLOWER PRODUCTION IN CHINA

China's cut flower industry began in Beijing, Shanghai and Guangdong in 1984. During the next 12 years, cut flower production grew steadily, and developed faster after 1990. The main reasons for this growth are: a) the consumption of cut flowers which has increased due to an increase in the standard of living and the rapid development of the tourist industry; and b) growers can gain relatively high profits from growing cut flowers. For example, the annual profit is US$18,750-65,625 per ha from cut flower cultivation, as compared to US$11,250 per ha from vegetable cultivation. Currently, cut flowers are grown almost all over the country, from Hainan in the South to Heilongjiang in the North, and from Shandong in the East to Xinjiang in the West. From a small start in 1987, commercial cut flower production in Yunnan Province grew at a higher speed than in any other area in China, moving this Province into number one position.

There are more than 30 cut flower species which are commercially grown in China, including Alstroemeria aurantiaca, Anemone, Antirrhinum, Asparagus, Anthurium, Calendula officinalis, Calla, Callistephus chinensis, Carnation, Centaurea cyanus, Chimonanthus praecox, Chrysanthemum, Dianthus barbatus,

Freesia, Gladiolus, Gypsophila, Gerbera, Heliconia, Hippeastrum, Lilium, Lisianthus, Limonium, Moluccella, Orchids, Peony, Pulsatilla chinensis, Prunus persica, Prunus mume, Ranunculus, Rose, Salix leucopithecia, Solidago, Sunflower, Tagetes patula, and Tulip. The growers can be divided into 5 categories: a) state farms or companies; b) collective farms; c) private farms; d) sino-foreign country joint ventures; e) foreign companies. Due to the nature of constantly expanding activities, it is difficult to obtain official statistical figures regarding China's cut flower industry. The data shown in this paper only partially represent the real situation.

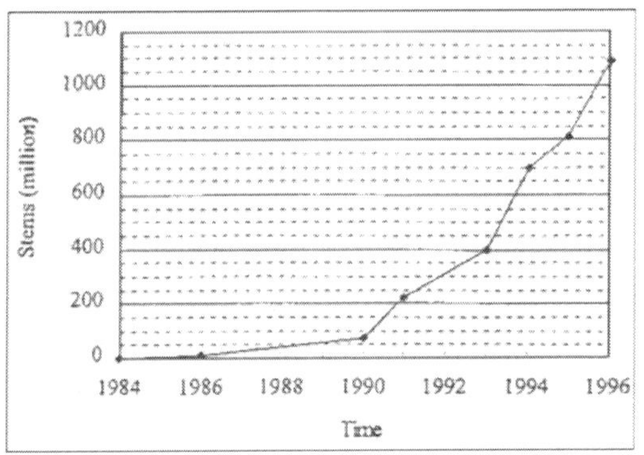

Fig. The Commercial Production of Cut Flowers in China During the Period from 1984 to 1996.

PRESENT SITUATION OF CUT FLOWER PRODUCTION

In 1996, the production area under cut flowers was about 3,000 ha and the yield was 1.09 billion stems. The major areas where cut flowers are grown commercially are Yunnan, Shanghai, Sichuan, Zhejiang, Guangdong, and Beijing. The production-acreage in Yunnan, Shanghai, Zhejiang and Beijing accounts for about 50 per cent of the total acreage in China. The annual production of cut flowers in these four regions accounts for about 79 per cent of the total production in the country.

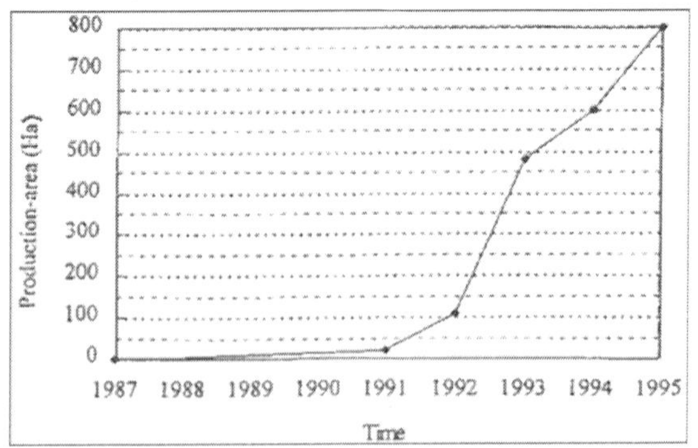

Fig. The Production-Acreage Devoted to Growing Cut Flowers in Yunnan Province During the Period from 1987 to 1995.

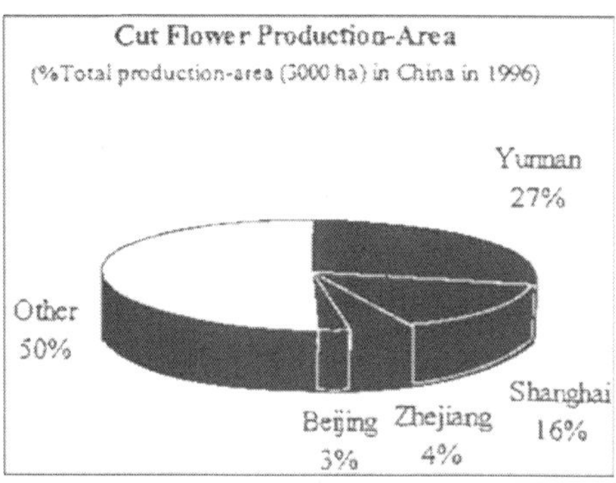

Fig. The Production-Area Devoted to Growing Cut Flowers in China During the Period from 1987 to 1995.

The top ten cut flower crops in China are rose, carnation, chrysanthemum, gladiolus, calla, gypsophila, anthurium, gerbera, lily and limonium. Roses are mainly grown in Beijing, Guangdong, Shanghai and Yunnan. In 1994 about 120 million stems of rose cut flowers were produced, accounting for about 17 per cent of all cut flowers. In 1995 the production-acreage of rose cut flowers was 300 ha, of which 100 ha were unprotected, and the annual production increased up to 200 million stems, accounting for about 18 per cent of all the cut flowers produced in the country. The major cultivars of rose cut flowers include Samantha, Red Success, Madelon, Kardinal, Carl Red, Dallas, Baccara, Rouge Meilland, Gabriella, Golden Emblem, Gold Medaillon, Golden Times, Gold Medale, Cocktail 80, Blami, Diplomat, Flamingo, Prima Donna, Sonia Meilland, Leading Lady, Bridal Pink, White Success, Tineke, Athena, Bridal White, Carte Blanche, Marina and Mercedes.

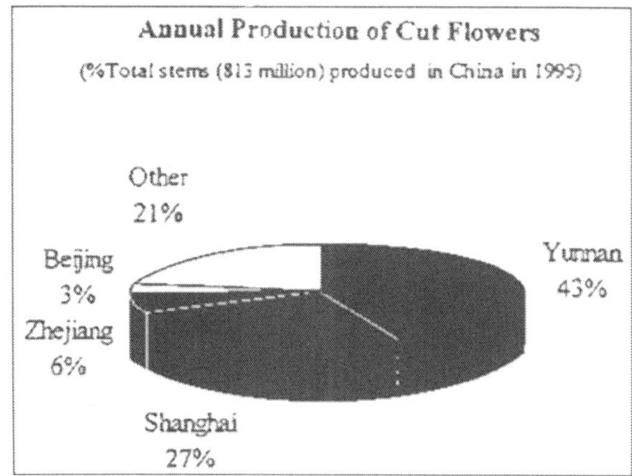

Fig. The Commercial Production of Cut Flowers in China in 1995.

Carnation prevalently of the standard type, accounts for about 25 per cent of all the cut flowers sold in the Beijing and Kunming wholesale markets. The production areas center in Yunnan and Shanghai. The major cultivars of carnation are from Israel, the Netherlands and Germany, such as Tasman,

Francesco, Omaggio, Red Rimon, Presto, Ondina, Dona, Red Corso, Hermes, Pink Francesco, Pallas, Napoleon, Sugar Lee, Rimo and Camba. The spray carnations are not popular in China because of consumer's preference for standard carnations.

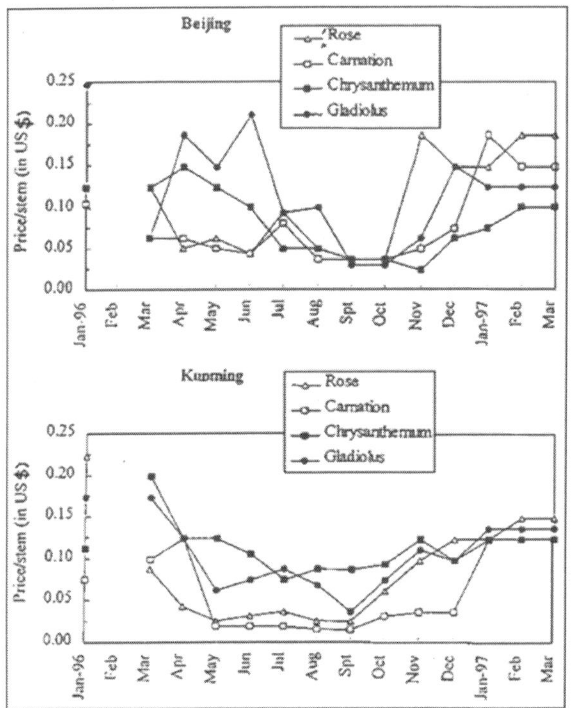

Fig. The Wholesale Prices of Rose, Carnation, Chrysanthemum and Gladiolus in Beijing and Kunming During the Period from January 1996 to March 1997.

Chrysanthemum accounts for about 20 per cent of all the cut flowers sold in the Beijing and Kunming wholesale markets. The production areas are concentrated in Guangdong, Shanghai and Beijing. The major cultivars of chrysanthemum include late-spring-flowering, summer-flowering, autumn-flowering and early-winter-flowering cultivars. Most of them are standard chrysanthemums. At present the spray chrysanthemums are not popular in China because of consumer's preference for standard chrysanthemums. Among various cultivars, yellow is the most appreciated colour, followed by white and red.

Gypsophila accounts for about 20 per cent of all the cut flowers sold in the Kunming wholesale market. The production areas are concentrated in Yunnan and Shanghai. There are semi-double and double cultivars with white flowers.

Gladiolus is mainly grown in Liaoning. There are 28 cultivars from Jining Province, 11 from Gansu Province, 11 from Hubei Province, 9 from Liaoning Province, and 26 from the Netherlands. Lily is mainly grown in Gansu. The major cultivars of lily cut flowers include Navona, Acapulco, Elite, Lorina,

Solemio, Pollyanna, Adelina, Akita, Her Grace, Jessica, Maremma, Amanda, Ankra, Apropas, Merostar, Wisdom, Snow Queen and White Satin. Gerbera is mainly grown in Yunnan and Shanghai. The major cultivars of Gerbera include Laurentius, Terranero, Michelle, Feugo, Shanghai, Estelle, Terramor, Clementine, Terraparva and Terratuba. Anthurium is mainly grown in Shenzhen and Beijing. The major cultivars of Anthurium include Alexia, Anneke, Candia, Gino, Gloria, Lydia, Margaretha, Mauricia, Nette and Rosetta. Calla is mainly grown in Shenzhen and Beijing. There are only white flower cultivars. Limonium is mainly grown in Yunnan, Sichuan, Guangdong and Beijing.

Cut flower production is year-round in various areas, and is usually under protected conditions except Hainan, Guangdong, Guangxi and Fujian where there are more cut flowers produced in open fields. The acreage of cut flower production under protected conditions is larger than that in open fields, but accurate data about the ratio of production between protected conditions and open fields are not available. Generally, during fall-spring, cut flowers are produced in greenhouses; and during spring-summer, they are produced out doors. Tunnels covered with plastic are widely used in growing cut flowers throughout the country. Modern greenhouses are also used, especially in Northern areas. In the last few years, new greenhouses and modern equipment have been introduced into China from Israel, the Netherlands and France.

In recent years, the growers in Beijing, Shanghai and Kunming have started using soiless culture and drip irrigation system to grow cut flowers. Controlled-release fertilizers have been used in Beijing for cut flower production. In the summer season, shade is widely used to reduce light to optimum levels for cut flower production. Artificial light has been applied by a limited number of growers in some regions such as Beijing and Shanghai to promote flowering of gladiolus, gypsophila, and lily during the winter, or to regulate flowering of chrysanthemum for year-round production. The technique of forcing bulbs at optimum temperatures is also practiced.

PRODUCTION OF PLANTING MATERIAL

Most of the cuttings, tissue culture plants and grafted planting material are produced in Shanghai, Yunnan, Beijing and Sichuan. Some of the propagation materials are virus-free. Although the companies in Liaoning, Jining, Beijing, Shanghai, Shaanxi and Shandong have produced some bulbs, most of the bulbs necessary for growing cut flowers in China are from the Netherlands.

The Beijing Flower Seedling Company which owns 5,000 square meters of glasshouse and 400 square meters of modern factory with tissue culture facilities, can produce 3 million plantlets every year, including rose, carnation, eustoma, chrysanthemum, lily, gerbera, etc. The North Star Garden Plants Company in Beijing can produce 1 million plants of anthurium, gerbera, Phalaenopsis amabilis and lily per year by means of tissue culture under automatically-controlled greenhouses (temperature and mist). Fifty thousand bulbs of lily are produced in Changping, Beijing, annually.

Each year, the Shanghai Floriculture Experimental Farm supplies 2.5 million carnation and gypsophila cuttings, gerbera plantlets and grafted rose plants, and 1.5 million bulbs of lily, gladiolus, iris and freesia. The Shanghai Sino-dutch Flower Co. Ltd. produces 1 million planting materials of gerbera and carnation annually. Recently a newly-built company equipped with computer-controlled facilities has been established in Yunnan that can produce 30 million planting materials annually. Another newly-built company in Kunming has a production capacity of 10 million carnation cuttings. A company in Liaoning Province supplies propagules year-round, including 11 million bulbs of gladiolus and 0.2 million bulbs of lily. A company in Sichuan Province has the capacity to produce 1.5 million bulbs and 1.2 million plantlets annually.

MARKETING

More than 90 per cent of the cut flowers produced in China are consumed within China. The cut flowers are sold through whole sale and retail distribution channels. At present, there are 670 cut flower markets and 7,000 flower shops across the country, and 7 regional whole sale markets which have been established in Beijing, Shanghai, Kunming, Guangzhou, Fujian, Chengdu and Liaoning. The major consumption areas are Shanghai, Beijing, Zhejiang and Guangdong. The majority of the flower shops are located in these areas. In 1996, 260 million stems of cut flowers were consumed with about 20 stems per capita in Shanghai, and 100 million stems were consumed with about 10 stems per capita in Beijing.

It is inferred from the analysis of the whole-sale prices of rose, carnation, chrysanthemum and gladiolus in the Beijing and Kunming markets during the period from January 1996 to March 1997 that these cut flowers were cheaper during the period from May to October than during the period from November to April. It is also obvious that the average wholesale price in Beijing is higher than that in Kunming. Yunnan supplies fresh cut flowers to more than 20 other regions such as Beijing, Guangzhou and Shanghai.

Table. Flower Shop Distribution in China.

Regions	Number of Flower Shops	% Total
China	7000	100
Zhejiang	2000	29
Shanghai	1300	19
Shenzhen	500	7
Beijing	500	7
Guangzhou	400	6
Hebei	200	3
Dalian	150	2
Hefei	107	1

Shenyang	100	1
Hainan	30	0.4
Guiyang	30	0.4
Jiangxi	20	0.3
Other regions	1663	23.9

The export of cut flowers to other countries mainly Hongkong, Japan, Singapore and Thailand, is very limited, less than 10 per cent of all the production in China. In 1993, China exported cut flowers of about US$1 million. In 1994, Yunnan exported cut flowers of US$4 million. In 1996, Beijing exported cut flowers worth US$750,000 (about 3 million stems through the Beijing Flower Service Cooperation), and Kunming exported more than 18 million stems.

POTENTIAL FOR CUT FLOWER PRODUCTION DEVELOPMENT

China has an abundance of wild and garden plants which can be developed into cut flowers. For example, Lilac and Forsytia are good early-spring cut flowers, but they are used only as garden plants. There are rich plant germplasms (15 species) of peony in China. Recently, China and Australia have done joint research to develop cut flower cultivars and the related cultivation and postharvest techniques.

There exist three types of regions suited to growing cut flowers in China: a) Dalian in Liaoning Province, and Weihai and Qingdao in Shandong Province, where the climatic conditions are similar to those in the Netherlands, with a very short day and freezing weather in winter and a very long day and warm weather in summer. With the appropriate technologies, it is possible to produce in Dalian, Weihai and Qingdao all the cut flower crops which are grown in the Netherlands; b) the eternal spring region like Kunming in Yunnan Province and Xichang in Sichuan Province, where the climatic conditions are similar to those in the Bogota plain in Colombia, with about the same length of day and night, and cool nights and warm days almost without change throughout the year; with the appropriate technologies, it is possible to produce in the Kunming and Xichang region all the cut flowers which are grown in the Bogota plain; c) the tropical region like Hainan Province and Xishuangbana in Yunnan Province, where the climatic conditions are similar to those in the Caribbean Islands and Thailand, with about equal length of day and night throughout the year; with the appropriate technologies, it is possible to produce in the Hainan and Xishuangbana regions all the cut flowers which are grown in the Caribbean Islands and Thailand. These three types of regions give China a unique potential to produce commercially all the known cut flower crops throughout the year.

China has a growing strength of labour because of the agriculture modernization and the increase in population. Currently, there are about 100 million extra labourers from the rural areas which can be used for growing cut flowers. The cost of labour is very low compared to the developed countries.

Presently China produces 1.09 billion stems of cut flowers annually. This represents a consumption of about 1 cut flower per capita per year, a very low level compared to other countries. More than 1.2 billion people are living in China. An increase of one cut flower in per capita consumption means an increase of more than 1.2 billion stems in the China's cut flower market. The people in China have a natural love for flowers, which is expressed by the care of gardens and public parks. We believe that with the improvements in the economic situation, the cut flower industry will expand rapidly.

Both the Yunnan and Guangdong Provincial Governments have evolved a series of policies to promote the development of cut flower industry. For example, Yunnan Provincial Government and local governments invested US$5 million in the cut flower industry annually in the last few years. They also reduced the cost of air freight and charged a low tax, only 2 per cent of the gate value.

CONSTRAINTS IN CUT FLOWER PRODUCTION DEVELOPMENT

Production areas are concentrated in the rural environments, but the consumption is concentrated in the cities. There is little direct communication between growers and consumers. Presently, the yield of cut flowers per square meter in only 50-80 stems in China, a very low level compared to other countries.

The quality of most cut flowers produced in China is only equivalent to that of the third grade in the international markets. The main reasons for the low yield and poor quality are: a) lack of specific knowledge and technology; for example, most growers who originally grew vegetables still treat flower crops as vegetables, using the local soil as growing media and traditional methods of irrigation and fertilization; in most growing areas, the local soil is heavy, unsuitable for continuous cultivation under the regime of heavy fertigation which tends to destroy the soil structure; the mist and drip irrigation systems are seldom used. b) lack of financial source to modernize the cultivation facilities. c) lack of disease-free planting materials; in order to save money, many growers use recycled materials for propagation instead of buying commercial planting materials. d) lack of cool chain from grower to retailer; treatments by chemicals for increasing vase life is not practiced extensively; the quality of cut flowers produced in Yunnan is reduced after they are transported to other areas such as Beijing and Guangzhou; until now there has been no quality standards for plant materials and fresh cut flowers.

The supply of cut flowers is centered around spring and autumn seasons, but during the Chinese lunar calendar new year Spring Festival when the consumption is the highest in the year, the supply is the smallest.

China has not joined the international organization UPOV yet. Therefore, it is difficult for the Chinese cut flower industry to obtain the newest and best cultivars bred by the breeders in other countries. There is an urgent need to pass the Plant Breeders' rights Law in China. This process has been started and China will have this law in the near future. This law is expected to promote investments in breeding efforts within China.

Conclusions

Cut flower production in China is being industrialized. The Chinese government has made "The Ninth Five-Year Plan from 1996-2000 and the Long-Term Objective Programme for 2010 for the improvement of cut flower production" as follows:

- The annual cut flower production will reach 2 and 4 billion stems by the year 2000 and 2010, respectively, in the country.
- The average per capita consumption in the urban areas will reach 10 stems by the year 2000.
- Cut flower production will be concentrated in Beijing, Shanghai, Guangdong, Yunnan, Sichuan and Hebei.
- The production of planting materials will be concentrated in Sichuan, Yunnan, Shanghai, Liaoning, Shaanxi and Gansu.
- Modern cut flower trade centres will be built in Beijing, Shanghai, Guangzhou, Kunming, Chengdu and Shenzhen.
- Optimum-scale cut flower whole-sale markets will be built in metropolitan centres across the country.

CUT FLOWER PRODUCTION IN INDONESIA

Indonesia is the world's largest archipelago of more than 13,000 islands with a total coastline of 81,000 kilometers. With a population of 190 million, Indonesia is the world's fourth most populous nation after China, India and the United States.

The country is in the process of accelerating the development of its economy through industrialization on the basis of self reliance, including rural economy. This will create employment opportunities and improve the welfare of millions of people living in the rural areas, as well as increase production quantitatively and qualitatively to meet the growing demand of the rapidly expanding population. This will offer strong investment opportunities especially for floriculture and other horticultural crops.

Ornamental horticulture, especially cut flower production has sprouted from cultural and hobby based activities into a lively and prospective business. Current government efforts to monitor the domestic and global demand and production has provided sufficient evidence that the cut flower industry could contribute substantially to the GDP of Indonesia. A productive cut flower industry could provide additional economic strength into the process of industrialization in the 21st century.

The government policy on horticulture development is directed to:

- Stimulate investment and create opportunity in the horticultural sector, based on agribusiness orientation;
- Reduce the sharp price fluctuation in order to maintain economic stability;
- Reduce import and increase export, to raise foreign exchange earnings;
- Create job opportunities and increase community income;
- Fulfill the demand for beauty, harmony and natural environment.

In the sixth Five Year Development Programme (Pelita VI), expected contribution from horticultural commodities to Product Domestic Bruto is 6.1 percent and growth of horticultural production is 5 percent per year . Approaches in the Second Long Term Development Plan for the horticultural sector will be focused on agribusiness orientation which is modern, economically efficient and sustainable.

PRESENT SITUATION OF CUT FLOWER PRODUCTION

Production centers of cut flower and ornamental plants have mostly developed on the basis of climatological and soil conditions and their distance from larger cities.

Main production centres at present are:

a. *Jabotabek:* Jakarta, Bogor, Tanggerang and Bekasi
b. *North Sumatera:* Brastagi, Kabanjahe, Tanjung, Morawa and Medan
c. *Riau:* Riau Islands
d. *West Java:* Cipanas, Cianjur, Bogor, Cisarua/Lembang, Sukabumi, Tasik Malaya

Due to the lack of statistical information, it is difficult to obtain figures about total area cultivated, production and productivity. For development purposes it is necessary to keep track of these data as is done for other horticultural crops.

The most important cities with regard to existing flower and ornamental sales as well as growing of orchids and ornamental plants are:

- *Java:* Jakarta, Bandong, Semarang, Malang and Surabaya
- *Sumatera:* Medan
- *Bali:* Denpasar
- *Sulawesi:* Ujung Pandang

The main features of the production centers for flowers and ornamental plants in Indonesia are described here under:

Jakarta (Jabotabek)

In general, every large city has its own orchid gardens which supply its own consumers. In Jakarta however, due to a scarcity of land and high price, orchid growers make investments outside Jakarta, for example in Tanggerang, Bogor and Bekasi.

Estimates of a 1993 census by DKI Jakarta reveal that a total of 102 hectares were cultivated under orchids and ornamental plants. Of this, 73 percent or 75 ha were in use for orchids. Orchid genera mostly grown by Jakarta growers are Dendrobium, Vanda, Arachnis, Oncidium, Phalaenopsis and Cattleya.

Major production areas of orchids in Jakarta are:

- *West Jakarta:* Kebon Jeruk, Grogol and Cengkareng
- South Jakarta: Kebayoran Lama, Cilandak, Lebak Bulus and Pasar Minggu
- *East Jakarta:* Jatinegara, Kramat Jati, Pulo Gadung, Pasar Rebo and Pondok Gede

Although orchid growing was primarily done in small shade net base gardens, presently some large scale orchid farms are established in Jakarta and its surroundings. Orchid growing is capital intensive compared to other cut flowers. An orchid farm of 1000 square meters will require a working capital of around US$ 12,000 to 15,000.

North Sumatera (Berastagi, Kabanjahe, Tanjung Morawa) and Riau Island

Production areas in Kabanjahe are scattered over 4 districts: Kabanjahe, Barusjahe, Tigapanah and Simpang Empat. Tanjung Morawa is located 17 km from Medan and is known as the center of production of ornamental plants in North Sumatera. Production of cut flowers include gladiolus, chrysanthemum, aster, gerbera, dahlia, anthurium and sedap malam (Polyanthes tuberosa). A few varieties of lilies and rose are also found. Farm sizes for flower growing in North Sumatera varies from less than 1000 square meters to one hectare. Most of the flowers are grown in combination with vegetables. Compared to Java less workers are used. Dry fish waste is used as fertilizer.

Gladiolus is planted in a phased manner so that harvest can be done continuously. In some fields, old plants are left for bulb production; generally, yellow, pink, red and dark red varieties are popular. Chrysanthemum is grown in open fields; white chrysanthemum is grown and pinched to yield one big flower, whereas in yellow and red varieties up to three flowering buds are left; rejuvenation is done after 18 months. Aster, gerbera and dahlia are more widely grown in North Sumatera than chrysanthemum. Some flowers are also sold without stem.

Most farms at Tanjung Morawa concentrate on more or less the same kind of ornamental plants, such as palms, pines, ixora, dieffenbachia, dracaena, cycas and croton. Plants are transplanted from the nursery into black plastic bags.

The wholesale market is located at Berastagi, especially on Tuesdays and Fridays, where flowers are traded and distributed to different markets and flower shops in Medan. Cut flower growers in Riau island produce Heliconia, one of the most prospective and commercial tropical flowers.

West Java (Cipanas, Bogor Cisarua/Lembang and Sukabumi)

Cipanas has been well known as a center for flower and ornamental production for a long time. Due to its favourable location and climate many inhabitants of Jakarta visit Cipanas on the weekends. Ornamental plants are sold directly to consumers alongside the roads. Major cut flowers produced are chrysanthemum, rose, gladiolus, gerbera etc. Chrysanthemum from Cipanas is well known for its high quality. In rose, mostly local varieties are grown. Cut flowers are supplied to Jakarta using bamboo boxes with banana leaves and are transported to hotels, florist shops and Rawa Belong market in Jakarta. Ornamental plants produced at Cipanas are pines, adiantum, azaleas, dieffenbachia, dracaena, bougainvillea, hibiscus etc. which are produced in small pots by villagers and sold alongside roads.

Bogor is known for its heavy rainfall with an estimated 320 days of rainfall in a year. Bogor has a very good connection with Jakarta and can be reached in half an hour by road. Cut flowers such as carnation, chrysanthemum, gerbera and cala lily are produced in plastic greenhouses at Megamendung and Ciawi. Ornamental plants such as palms, pines and adiantum are produced at Ciapus Bogor. In Lembang cut flowers are grown on a small scale and are less important than ornamental plants. At Cisarua, a desa (village) next to Lembang, the villagers are producing ornamental plants in almost every home garden. It is estimated that there are around 100 growers with an average farm size of 500 to 1000 square meters. Plants are grown in plastic bags in a medium of rice hulls mixed with stable manure. They mainly consist of flowering pot plants such as baby roses, asters, mini carnation etc. Young plants of pines, cycas, azaleas, dracaena and palms are also produced.

Cut flowers such as gladiolus, anthurium and gerbera are mostly produced at Selabintana and Sukabumi. Sedap Malam (Polyanthes tuberosa) is produced at Selabintana/Sukabumi, Mayak/Cianjur and Indihiang/Tasik Malaya. Area under tuberose in Sukabumi is 6 hectares, in Cianjur 39.15 hectares and in Tasikmalaya 9 hectares. From the wholesale market the flowers are transported to Jakarta.

Central Java (Bandungan, Tegal, Pemalang, Purbalingga and Magelang)

Bandungan is the main production area for cut flowers in Central Java. Most of the flowers at Bandungan are produced on small farms by a large number of farmers in a mixed cropping system with vegetables. From an agricultural census conducted in 1986 it appeared that there were 1530 farmers, who produced both flowers and vegetables. Thirty six of them had an area larger than 0.5 hectare. Bandungan produces cut flowers of aster, roses, tagetes, gerbera, dahlia, gladiolus, chrysanthemum, carnation, lily and amaryllis. Quality of flowers varies a lot and grading is done by the trader.

Tegal and Pemalang produces jasmine (Jasminum sambac). It is used as an ingredient for the taste of green tea. Banjarnegara has the largest jasmine plantation area of 345 hectares while Banyumas has 45 hectares. Sedap malam (Polyanthes tuberosa) is produced by growers in Central Java at Desa Citrosono of Magelang and Bandungan. Harvested area under tuberose at Citrosono-Megelang is around 14.5 hectares and at Bandungan-Semarang 12.3 hectares.

East Java (Batu, Pujon, Tretes, Pasuruan and Madura)

Batu is the center of cut flower and ornamental plant production in East Java. Batu and Pujon, are located at high altitudes of 1000 to 1300 meters above sea level. Most of the farmers at Batu and Pujon use mixed cropping patterns of flowers and vegetables with an average of 0.5 hectare per farmer. Cut flowers produced are gladiolus, rose, chrysanthemum, orchids, gerbera, lilies and tagetes. In Tretes ornamental plant growers have formed an association called Aspeni (Asosiasi Pengusaha dan Petani Flora Indonesia). Ornamental plants produced at the 3 locations above are pines, palm, cycas, ficus, ixora, dracaena, azalea and dieffenbachia.

Pasuruan is the center of sedap malam (tuberose) production in East Java. Total harvested area of tuberose is 75 hectares at Bangil and 69 hectares at Rembang, which are sub-districts of Pasuruan. Growers in Pasuruan produce jasmine beside sedap malam. Harvested area under jasmine in Pasuruan is 15 hectares. Madura is one of the most famous areas for jasmine production in East Java. Area under jasmine at Bangkalan-Madura is 30 hectares. Most jasmine production in East Java is used for accessories at wedding ceremony and other religious activities. The local wholesale market of cut flowers is located at Batu. Some of the growers act as traders as well and packing of flowers is done at their home.

TECHNOLOGY GENERATION

Research in floriculture is carried out by the Research Institute for Ornamental Plant (RIOP) in Jakarta which has had its mandate from the Ministry of Agriculture since 1995. RIOP has two installations at Cipanas and Segunung. Major areas of research are breeding, agronomy, pest and disease control, post-harvest technology and economics. Attention has been directed to cut flowers namely orchids, rose, jasmine and tuberose.

The on-going research programmes within the RIOP are as follows:

- Germplasm collection, characterization, elevation and conservation.
- Quality improvement on cut flower priority and ornamental plants.
- Growth media and plant nutrition.
- Cropping system including hydroponics.
- Pests and disease control of cut flower priority.
- Vaselife prolongation of cut flower.
- Socio-economic studies, to increase grower income and marketing.
- Seed technology and planting material production.

Research on orchids mostly relates to quality improvement, virus resistance, comparison of growth media and plant nutrition from the tissue culture stage through transplanting and cut flower production. Ways of rapid multiplication of orchid plants with tissue culture have been studied as well. Types of orchids include Dendrobium, Phalaenopsis and Vanda. On roses work was done on variety improvement, adaptability testing, planting material production and socio-economic aspects. Prolongation of vaselife of cut flowers and colouring tuberose by testing several formulas have also been studied. A germplasm collection of cut flowers and ornamental plants is maintained at the Research Institute for Ornamental Plant.

Impact of research result is clear through interaction between growers and the research institute. Moreover, dissemination of technology could be further enhanced by seminars and publications. At present there is an institution called BPTP which has the mandate for technology assessment, development and extension work in almost every region.

PRODUCTION OF PLANNING MATERIAL

At present many varieties are being grown in Indonesia both local as well as imported. It needs to be determined which ones are more suitable for local growing conditions and match consumer preferences. With the rapid development of the local market of cut flowers and ornamental plants, larger quantities of high quality planting material will be required. Specialization on growing planting material of a certain kind of cut flowers and growth stage of ornamental plants is promising and looks prospectful. Multiplication by means of tissue culture of cut flowers and ornamental plants has been found promising because of the rapidity of producing a large number of plantlets in short time, and the uniformity of planting material which is free from diseases. Propagation of ornamental plants is done in partial greenhouses which provide shading and protection against rain and pests. Simple overhead sprinkler irrigation system provides easy water supply and control of humidity.

Table. Estimation of Cut Flower Planting Material Demand, Import and Production

Crop	Seedling Requirement	Import	Seedling Production
Orchid	66,200,000	1,100,000	65,100,000
Rose	5,600,000	-	5,600,000
Chrysanthemum	76,400,000	100,000	76,300,000
Anthurium	10,800,000	400,000	10,400,000
Carnation	28,800,000	500,000	28,300,000
Lilium	32,400,000	100,000	32,300,000

Planting material production in Indonesia is not yet developed. There is no company breeding for cut flowers at present in Indonesia. The market segment which uses high quality planting material is small but shows a growth of about 10 percent per year.

MARKETING

In regard to the market for cut flowers in Indonesia, it must be noted that no integrated domestic market exists. The market consists of a number of confined markets, larger cities and their supply regions. A distinction should be made between Jakarta and other larger cities in Indonesia. In Jakarta giving flowers for all social occasions, such as birthdays, weddings, illness, business attention etc. has become rather common and substitutes for personal visits. As a result of this, Jakarta has become by far the largest market for flowers in Indonesia and consumes approximately 76 percent of the total consumption. In Jakarta more than 900,000 stems of cut flowers are sold per week, accounting for an approximate value of US$ 5.1 million per year, whereas the total amount for all major cities is estimated as US$ 6.8 million per year.

Table. Consumption of Cut Flowers in Jakarta (Stems).

Year	Rose	Chrysant.	Tuberose	Gladiolus	Anthurium	Dahlia	Others	Total
1993	8,038,800	7,035,000	6,156,800	7,543,500	614,300	2,473,100	3,078,400	33,939,900
1994	8,823,300	7,718,700	6,753,500	7,180,700	673,800	2,714,300	3,376,700	37,241,000
1995	9,682,300	8,466,700	7,406,000	8,878,100	739,000	2,978,400	3,703,000	40,853,500
1996	10,622,400	9,384,700	8,119,200	8,641,100	810,100	3,267,300	4,059,600	44,804,400
1997*	11,650,800	10,178,700	8,898,200	9,475,400	887,900	3,583,400	4,449,100	49,123,500
1998*	12,775,100	11,155,100	9,748,400	10,387,000	972,700	3,928,900	4,874,200	53,841,400
1999*	14,003,700	12,220,800	10,675,700	11,382,500	1,065,200	4,306,300	5,337,900	58,992,100

POTENTIAL FOR CUT FLOWER PRODUCTION DEVELOPMENT

Tissue culture technology for multiplication of plantlets is rapidly developing and has been accepted on a wide scale for several flowers and ornamental plants. Advantages lie in less time needed, uniformity of produced plantlets and virus free planting material.

Indonesia has ideal climatological conditions for growing tropical plants. Presently an interesting domestic market has developed, where relatively high prices are fetched, based on the scarcity of the product more than the actual cost price.

To be able to achieve the quality standards required in the international market and its effective marketing, cooperation with international growers is a feasible alternative to obtain results in the not too distant future for Indonesian growers. In order to enter the world market to some significant extent Indonesia needs to start selection and breeding of its own varieties. A substitute for soil mixed with rice husk as a growth media for ornamental plants is required in case of export since most countries prohibit soil and rice husk to enter their territories.

CONSTRAINTS IN CUT FLOWER PRODUCTION DEVELOPMENT

a. Although several commercial tissue culture laboratories are in operation in Indonesia, all have limited operations to serve their own requirement or a small circle of associated farmers. By doing so, the scale of operations remains too small for a profitable undertaking and consequently investments in motherstock of planting material are not made.

b. No specialized nursery for young ornamental plants is in existence. Growers are not able to follow the market trend quickly and multiplication is done in an inefficient way.
c. Indonesian private growers of ornamental plants and flowers were left on their own for variety testing and developing appropriate cultivation methods. This has resulted in high cost and slow spread of adapted varieties and cultivation techniques. Support from RIOP as technology generating institution and BPTP for transfer of technology is needed in every region.
d. Institutional promotion to increase popularity of flowers and ornamental plants in Indonesia is still limited. Similarly, information about flowers and plants which can be supplied from Indonesia is almost lacking in the world market.
e. Most of the cut flower production centers are scattered and located too far from the distribution point or air port.
f. High interest rate on credit for capital on agribusiness which is treated the same as other commercial investments.
g. Limited area/land at reasonable price.

Conclusions

Research on cut flowers and ornamental plants is very important to support and solve the problems faced by growers. The cooperation between private and public institutions needs to be strengthened and developed. The development of the domestic market is a prerequisite for the development of export which requires high standard quality and large volume as well as competitive price.

The development of a well established wholesale market is needed in Jakarta to have a positive impact on the development of this sector.

The short production cycle of cut flowers and ornamental plants, makes the country an ideal place for multiplication of planting material, which needs to be explored.

CUT FLOWER PRODUCTION IN MALAYSIA

The cut flower industry in Malaysia is a relatively recent development compared to other agricultural enterprises. From its rather humble origins as a hobby industry, the cut flower industry in Malaysia has developed into a very viable commercial enterprise with the most marked growth in the mid-Eighties. In fact it has shown such tremendous growth in the last decade that production has increased tenfold and export twelve-fold in response to local and foreign demands. The trend is expected to continue in the future with growing affluence of the local population and that of the developed countries as well as improved market opportunities. The cut flower market consists of 3 important components *viz.* temperate flowers, orchids and other lowland flowers. In general, the area of cultivation of cut flowers in Malaysia is determined by the climate and topography of the land. For instance, highlands such as the Cameron Highlands are the major growing areas of temperate flowers. Other cut flowers adapt better to the hot humid conditions in the lowlands with orchids constituting the major share of the production.

In the National Agriculture Policy (1992-2010) and the Seventh Malaysia Plan (1996-2000), cut flowers have been identified as a priority group of crops with good potential to meet the growing domestic and international demand and to generate higher income for producers.

The main thrusts of the policy are reflected in the following areas:

- Substantial expansion of flower production of both the lowland and highland varieties to meet the demand of an expanding market, in particular world demand.
- Cultivation of lowland varieties on marginal as well as former mining areas.
- Efficient and reliable supply of planting materials.
- Intensification of R and D in the development of better varieties, cost effective production technology and post harvest handling.
- Diversification of exports beyond the traditional markets.
- Encouragement of foreign investment to bring in more advanced technology and provide access to international markets.

It has been estimated that at least 1,000 farmers are involved in the floriculture industry that contributed RM 370 million to the economy in 1995. Orchids contributed 40 percent of total value of production, followed by temperate cut flowers (33 percent) and ornamental plants (27 percent).

Although the contribution of cut flowers to the total of Malaysian exports is not significant, the annual growth rate over the last year was 24 per cent. If the forecast of 10 per cent annual growth rate of floriculture production holds true for the 7th Malaysia Plan period (1996-2000), the floriculture industry will enjoy the highest annual growth rate compared to other major commodities including rubber and palm oil over the same period.

PRESENT STATUS OF CUT FLOWER PRODUCTION

Major Production Areas

The major production areas for temperate cut flowers are in the Cameron Highlands, Pahang State (638 ha), Gua Musang, Kelantan State (10.3 ha), and

Ranau, Sabah (85.1 ha). In the lowlands the main orchid cut flower production areas are located in Johore Bahru, Johore State (162.5 ha), Batang Padang, Perak State (52.4 ha), Kota Tinggi, Johore State (32.89) and Petaling, Selangor State (27.36 ha). Smaller quantities of orchids and other lowland cut flowers are also produced in the states of Melaka, Negeri Sembilan, Kelantan, Perlis and Sarawak.

Area Under Cultivation and Flower Types/Varieties

The total area under cut flowers is estimated to be over 1,218 hectares, out of which 580 hectares are under orchids and 638 hectares are under temperate cut flowers.

The major types of cut flowers and popular varieties of each species grown commercially are as shown in Tables 1 and 2. In terms of number of flower stalks, temperate flowers contribute 71.46 per cent towards the total cut flower production while the remaining 28.54 per cent is from orchids.

The most popular orchid types cultivated are Dendrobium (11.17 per cent), Aranda (8.0 per cent), Oncidium (5.01 per cent) and Mokara (3.5 per cent). Together these 4 orchid genera produced 97 per cent of the country's orchid cut flowers in 1994. The most important temperate flower types are rose (33.48 per cent), chrysanthemum (22.62 per cent) and carnation (9.02 per cent). These three flower types contributed 91.1 per cent of total temperate cut flower production.

OPEN FIELD AND PROTECTED PRODUCTION SYSTEMS

The type of production system used for different species of cut flowers in Malaysia is dependent on weather conditions, cultural requirements and quality maintenance, as described below:

Temperate Flowers

The climate of the Cameron Highlands, where the mean day and night temperatures are 14°C and 22°C, respectively, is very conducive to the cultivation of a wide range of temperate cut flower species.

However, the total annual rainfall of 2700 mm that falls mainly in April-May and October-December affects growth and vigour of the plants, increases disease incidence and pest infestation, reduces yield and quality and causes erosion of top soils. Thus, almost all temperate cut flowers such as rose, carnation, chrysanthemum, peacock and lilium are grown under simple plastic shelters.

These are simple open structures with height of 3 m and each consists of a wooden or galvanised iron frame with a polyethylene roof to protect the cut flowers from rain. Anthuriums, on the other hand are planted in shadenet houses with shading level of 75-80 per cent. Irrigation is generally provided by overhead sprinkler daily during the first month of growth and gradually reduced during later phases. Surface and drip irrigation is used when the flower starts to open. Processed organic fertilizers, compound fertilizers and foliar fertilizers are used at different stage of growth.

Chrysanthemum is different from the other flower types because there is a need for disbudding, application of plant growth regulator (to regulate height, strengthen stems and for quality blooms) and the use of artificial lighting and light interruption for vegetative development and floral initiation, respectively.

Table. Main Temperate Cut Flower Types and Varieties and Percentage Composition of Total Cut Flower Production (1994).

Main Flower Types	Main Varieties	% of Total Cut Flowers Produced
1. Rosa	Hybrid teas - Samantha, Etna, Sonia, Kiss, Dukat, Amber Green, Cocktail 80, Golden Times, Madelon	33.48
2. Chrysanthemum	Reagan Yellow, Reagan Dark Spendid, Jaguar Red, Vyron, White Spider, Fiji Pink, Cleopatra Royal, Yellow Standard, Red Japanese Standard, Rivalry, Ice Follies White	22.62
3. Carnation	Standards - Master, Killer, Oriana, Indios, Liberty, Nicol, Samor Spray - Elsy, Etna, Cherry Fantasia, Cartauche, Iceland	9.02
4. Aster (Peacock flower)	Monte Cassino, Suzanne, Pointed Lady, Solidago, Suntop	2.36
5. Gerbera	Melody, Beauty, Sundance, Mickey	1.77
6. Limonium	Misty Blue, Misty White	0.63
7. Anthurium	Nitta Orange, Anneke Pink, Avanti	0.60
8. Lilium	Cassablanca, Star Gaze, Dreamland, Harmony, Snow Queen, Olympic Star	0.44
9. Heliconia	Spaciata Purple, Spaciata White	0.30
10. Liatris		0.08
11. Solidago		0.05
12. Gladiolus		0.05
13. Strelitzia	Bird of Paradise	0.02
14. Gypsophila		0.02
15. Alstromeria		0.01
16. Dahlia		0.00
17. Helichrysum		0.00
Subtotal for Non Orchids		71.46
Subtotal for Orchids		28.54
Grand Total for Cut Flowers		100.00

Due to continuous cultivation on the same piece of land, the problems encountered by growers include salinity and the build-up of soil-born diseases and nematodes. Leaching of saline soil and the addition of more top soil are some methods practised by growers to overcome salinity problems. Expensive soil fumigation techniques using steam or chemical fumigants are used to control soil-borne diseases and pests.

Table. Main Orchid Cut Flower Types and Varieties and Percentage Composition of Total Cut Flower Production (1994).

Main Flower Types	Main Varieties	% of Total Cut Flowers Produced
Orchids		
1. Dendrobium	D. Sonia, D. Tomie Pink,	11.17
	D. Shavin White, D. Chanel,	
	D. Sharifah Fatimah, D. Tuang Pink	
2. Aranda	A. Wan Chark Kuan,	8.00
	A. Noorah Blue, Aranda Kooi Choo,	
	A. Happy Beauty, A. Anne Black	
3. Oncidium	O. Golden Shower,	5.01
	O. Gower Ramsay, O. Taka	
4. Mokara	M. Chark Kuan (Pink, Orange, Red),	3.50
	M. Dinah Shore, M. Khaw Phaik Suan	
5. Arachnis	A. Maggie Ooi	0.56
6. Aranthera	A. Anne Black, A. James Storei	0.13
7. Vanda	MAS Los Angeles	0.09
8. Holttumara	H. Loke Tuck Yip, H. Chee Kiong	0.06
9. Phalaenopsis	P. Sun Prince Pink,	0.01
	P. Hawaiian Clouds x P. Carmela Pink	
10. Kagawara	K. Christine Low	0.01
11. Cymbidium		0.00
Subtotal for Orchids		28.54

Orchids

The three major orchid types of Malaysia are grown under different production systems. The more popular sympodial orchids such as Dendrobium and Oncidium are grown under shadenet with shading level varying between 25-50 per cent depending on the species. For the production of quality white dendrobium blooms that are free of specks and other blemishes, these varieties are sometimes grown in plant houses with roofs of polyethylene or transparent artex. For those Dendrobium species that are grown in shadenet houses without polyethylene roof, excessive rains are known to cause bud drops and flower specks in light-coloured varieties.

Most sun-loving monopodial orchids such as vandaceous types are grown in open fields while some Mokara and Aranda hybrids are also cultivated in shadenet houses with 25 per cent shading.

ADVANCES IN PRODUCTION TECHNOLOGY

Varietal Improvement

Three programmes for varietal improvement of orchids are currently being carried out, namely conventional breeding, genetic engineering and mutation breeding. The germplasm collection has increased to 500 species representing 100 genera. Another 196 accessions are being maintained as "studs" for hybridization work. Conventional breeding has resulted in the release of at least 7 hybrids consisting of 4 Dendrobiums, 1 Vanda, 1 Cattleya and 1 Kagawara.

Research in genetic engineering has just been initiated to induce colour change in established varieties, for instance changing yellow Oncidium (O. Golden Shower and O. Gower Ramsay) to white. Meanwhile studies on in vitro mutagenesis include the use of mutagens and gamma ray irradiations on Aranda, Mokara, and Vanda orchids. Induced mutation research on Dendrobium has indicated variations in pigmentation and flower size in response to irradiation.

For temperate cut flowers, varietal evaluation trials have identified promising varieties of chrysanthemum, rose, carnation, lilium, anthurium and gerbera for highland cultivation. Induced mutation studies on 3 chrysanthemum varieties have also indicated that the recommended optimum dose for the mutagenesis of these varieties is 30 Gray.

Cultural Practices

In orchids, the application of fertilizers at night and in the morning helps to increase inflorescence yield. Oil palm bunch compost when mixed in charcoal medium improved seedling growth by three times in Oncidium, twice in Dendrobium and one and a half times in Aranda seedlings. The quality of chrysanthemum cut flower can be improved significantly through the application of calcium up to 150 mg/1. Cyclic light interruption as opposed to continuous night light interruption reduces energy cost and thus lowers the cost of chrysanthemum cut flower production.

Disease Control

In orchids, necrotic specks in Dendrobium blooms are common in plants grown under plastic shade. The spraying of 1 g/1 Ronilan every 10 days can effect complete control of this problem. The black nose disease of orchids has been confirmed to be caused by the association of Fusarium monoliforme with thrips. This disease can be controlled by alternate sprayings of Mesurol and Dithane M 45. Chemical control measures have also been developed for Sclerotium rot and pseudobulb rot, Cercospora leaf spots and alga.

The white rust problem on chrysanthemum that has restricted its import by countries such as Australia and Japan has been overcome with rust resistant varieties such as Tigerrane, Tiger, Target, Naru and Puma Yellow. Effective chemical control is also obtained with fungicides such as bromuconazole, bitertanol, flutolanil, myclobutanil, tridemenol and propiconazole. Powdery mildew in roses can also be controlled by chemicals such as sulphur and cupric oxide.

Pest Management

Leaf miners are a major pest problem of cut flowers in the Cameron Highlands. Studies on biological control have indicted that two eulophid parasitoid species (Hemiptarsenus varicornis and Chrysocharis sp.) parasitize both species of leaf miners (Liriomyza sp. and Chromatomyia horticola) found in a wide range of cut flowers. Meanwhile neem extracts have been found to be effective against leaf miner in potted chrysanthemum.

Studies have identified the most serious nematode pests as Eadopholus simillis on chrysanthemum and Pratylenchus coffeae on chrysanthemum and anthurium. These pests can be controlled by the application of nematicide such as fenamiphos, carbofuran, dazonmet and isazophos and adoption of preventive phytosanitary measures. Effective chemical control measures have also been developed for the beetle Oulema pectrolis and snail.

Post Harvest Handling

Studies indicated that the telescopic box type is suitable for packing temperate cut flowers while treatment of rose and carnation flowers at the right stage with certain flower preservatives has reduced bent-neck problem and extended its shelf-life. In addition, preservation techniques have been developed for rose, carnation, peacock, golden rod, anthurium, statice and cosmos for use as dried and pressed flowers. Although orchids are less perishable than temperate cut flowers, storage temperature does affect its quality. Dendrobiums can be stored at temperatures between 10-15°C without any sign of scorching.

All mites and thrips on roses, carnations, chrysanthemums and orchids are killed by methyl bromide fumigation without affecting the shelf life of the flowers. Evaluation of several chemicals as fumigants has yet to come out with an effective alternative to methyl bromide.

PRODUCTION OF PLANTING MATERIAL

A survey of the cut flower industry conducted in 1995 indicated that 107 orchid growers produced their own planting materials, 241 producers obtained their planting materials from local nurseries while 122 growers imported theirs from other countries. The corresponding figures for non-orchid cut flowers are 107 (own production), 294 (local) and 134 (import). Based on the application of import permits, import of planting materials during the first 6 months of 1996 consisted of 3.43 million cuttings/plants, 1.01 million bulbs/corms, 156 kg of flower seeds, 51,595 flask of tissue cultured orchid materials, 4,000 community pots of orchids, and 330,450 orchid seedlings.

Although many temperate cut flower growers produce planting materials, a large number purchase these from local nurseries or importers. In the case of chrysanthemum, the most widely grown temperate flower, two companies namely Van Der Kemp and Fides, produce most of the rooted and unrooted cuttings. Van Der Kemp has a 15-acre nursery that produces cuttings in imported Dutch greenhouses under drip and mist irrigation systems. Fides operates a smaller 10-acre nursery with cuttings produced under plastic rain shelters that are equipped with similar irrigation facilities. These two nurseries together produce an estimated 50 million chrysanthemum cuttings. About 80 per cent of these materials are supplied to growers in the Cameron Highlands, while the rest is exported to nearby ASEAN countries. It is estimated that a total of 300 million chrysanthemum cuttings is required by growers in the Cameron Highlands and both these companies can only produce between 10-13 per cent of the local needs. There are more than a dozen commercial tissue culture units producing orchid materials. These are mostly small operations producing either meristem culture or seed culture. Some major orchid growers also contract out the production of planting materials of orchid hybrids to large tissue culture laboratories in Thailand and Singapore.

MARKETING

Marketing of cut flowers in Malaysia is less organized than that in Western countries with different degrees of sophistication depending on flower varieties traded, farm size and destinations of the produce. The various marketing channels through which cut flowers.

The multi-tiered marketing and distribution system is rather complicated and according to growers does not pay premiums for good quality. Malaysia is studying the possibility of establishing an auction market. However, growers still play an important and direct role in the distribution to retailers in the local market as well as to importers in the foreign market. The grower can either deliver directly (large farms) or through an export agent (small farms) to the importer. The flowers are then sold directly to wholesalers or in auction markets (foreign countries) before distribution to the retail markets and from there to the consumers.

There are several associations of cut flower producers but these are formed to provide service and supplies to members only. The more significant ones are the Cameron Highlands Floriculturist Association and the Commercial Orchid Growers' Association of Malaysia. Although these associations do not collectively market their produce there is growing realization of the benefits of such marketing arrangement. Several temperate flower producers have already grouped together to form a trading company to deal directly with importers in foreign countries. Some companies also specialize only in export and buy directly from growers who have production contract with them.

The exports of floricultural products increased from RM 6.5 million in 1985 to RM 54 million in 1994 showing an eight-fold increase over a ten-year period. However, conservative official records show a decline in the export value of cut flowers from a high of RM 40.63 million in 1992 to a low of RM 18.31 million in 1995, although there was an increase in cultivated production area. This two conflicting trends appear misleading but may be explained by the low market prices that plague the industry in 1995, under declaration of import volume and value, movement to lower-price markets and increased domestic consumption. The main export markets are Japan, Singapore and Hongkong. Japan is the leading export market for orchids, Singapore for ornamentals and Hongkong for temperate flowers.

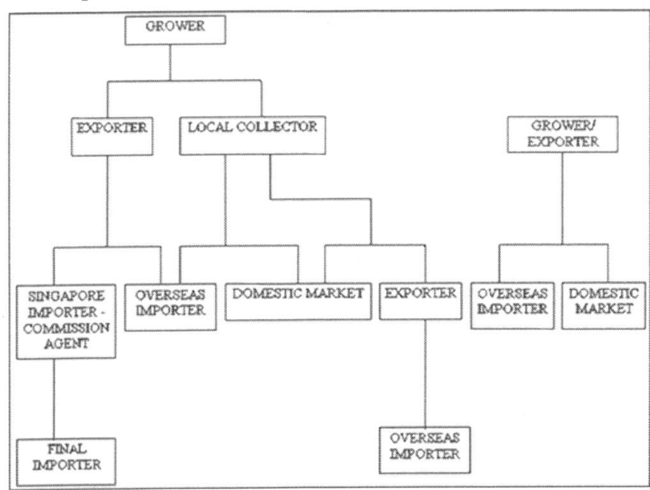

Fig. Marketing Channels for Flowers in Malaysia.

Import of cut flowers has shown a fourfold increase from RM 0.73 million in 1992 to RM 2.77 million in first eleven months of 1995. The main countries exporting cut flowers to Malaysia are the Netherlands, Singapore and Thailand.

POTENTIAL FOR CUT FLOWER PRODUCTION DEVELOPMENT

The prospects for the cut flower industry in Malaysia are very bright due to the growth in domestic and export markets. The domestic per capita consumption has increased from RM 0.71 in 1990 to its current level of RM 2.50. This trend is expected to increase further with greater consumer affluence and growing appreciation for fresh cut flowers. Other factors that contributed to this forecast is the rapid economic growth and the fast rate of urbanization in the country. There is also an increasing demand for cut flowers in the international market that is expected to continue at a growth rate of 6 per cent per annum. This will result in a corresponding increase in value that is expected to reach RM 20 billion and RM 36 billion in the years 2000 and 2010, respectively. There is tremendous scope for expansion of the export potential of Malaysian cut flowers, as major consuming countries are facing rising production cost and will increasingly depend on lower-cost producers to supply their requirements. Malaysia has several advantages including a wider product range, longer shelf-life for Malaysian orchids, potential to exploit its biodiversity for product development and ability to export flowers year round.

New markets especially for Malaysian orchids are emerging in the Middle East, Northern Europe and East Asia. Existing markets such as Belgium, Finland and Ireland are expected to expand further.

CONSTRAINTS IN CUT FLOWER PRODUCTION DEVELOPMENT

Although the cut flower industry in Malaysia has shown impressive growth in the last decade, it has not been able to realize its full potential because of several constraints. Among the more important ones are the following:

Table. Exports of Cut Flowers According to Types from 1992-1996 (RM '000).

Type/Year	1992	1993	1994	1995	1996*
Orchids	6,259.60	4,268.40	9,034.49	6,828.24	6,069.18
Temperate Flowers	33,305.84	20,450.93	23,283.37	9,632.21	9,140.85
Other Cut Flowers & Flower Buds, Other than Fresh	1,063.82	753.08	1,394.76	1,853.90	1,239.78
Total Export	40,629.26	25,472.41	33,712.62	18,314.35	16,449.81

Note:

* Data based on Jan-Nov. 1996

Table. Import of Cut Flowers According to Types from 1992-1996 (RM '000).

Type/Year	1992	1993	1994	1995	1996*
Orchids	114.17	14.28	39.03	74.88	977.10
Temperate Flowers	287.46	171.69	434.87	433.27	701.20
Other Cut Flowers & Flower Buds, Other than Fresh	326.03	392.55	461.15	747.47	1,093.30
Total Import	727.66	578.52	935.05	1,255.62	2,771.60

Note:

* Data based on Jan-Nov. 1996

Economic

The intense competition for land resources with more lucrative economic activities, high development cost, short land tenure and environmental considerations (highland production) impede the expansion of the cut flower industry both in the highlands and lowlands. This is further compounded by the high initial investment for planting materials, irrigation systems, shade and other infrastructure facilities. Financial institutions are also unwilling to provide credit for floriculture projects due to perceived high risks.

Technology

The development of new varieties has not been able to keep pace with changing consumer preferences. Moreover, inconsistency in the supply of uniform and high quality planting materials and the slow rate of technological advancement in critical areas such as post harvest handling, storage and transportation have resulted in export of lower quality produce with shorter shelf-life.

Infrastructure Support

Existing marketing practices do not ensure premium prices for good quality. Prices are also not normally fixed at the point of sale in the open account payment system. This often results in non payment, high arbitrary discounts and commissions especially by buyers from Singapore and Hongkong. Insufficient facilities for refrigerated transportation and storage have led to high post harvest losses and export of poor quality produce. Exports are also confined to traditional markets due to insufficient air cargo space and handling facilities at airports.

Institutional Issues

The high tariff barriers imposed by Taiwan, Thailand and Europe and the strict phytosanitary requirements enforced in Japan, the United States of America, Australia and New Zealand, are other impediments to further development of the industry. These problems are further compounded by the lack of a standard domestic grading and inspection system.

Human Resources

There is a general lack of skilled human resources in important areas such as production, post harvest handling, product handling, product development, biotechnology and auctioning. This can become a serious handicap in the highly competitive cut flower industry, where innovation and responsiveness to changes can give local growers the edge over their rivals.

CONCLUSIONS

The prospect for cut flowers is very bright taking into consideration the growing demand in the domestic and export markets and the fact that flowers are relatively recession-proof. The Malaysian Government has recognised this fact and has accorded priority to floriculture in the National Agriculture Policy, providing various tax and financial incentives to attract investments. Malaysia should make full use of its inherent advantages and attempt to rise above current constraints and provide an adequate and consistent supply of quality cut flowers during periods of maximum demand. Current production and marketing activities including development of desired varieties, and the search for new markets should be upgraded with the full cooperation of all parties concerned. These efforts and other future development plans are expected to enable Malaysia to make the quantum leap and become one of the leading producers of tropical cut flowers in the world.

A successful development plan will have to pay particular attention to the following areas:

- Expansion of domestic production through the efficient utilisation of domestic resources to exploit the opportunities that exist.
- Strengthening market and market capabilities through better cooperation between private sector and Government agencies in order to expand existing markets and penetrate new ones.
- Strengthening the institutional support and support services for continuous innovation and development of capital-intensive production

technologies to improve product quality and to meet changing consumer demand.
- Exploitation and enhancement of processing and downstream capabilities.

CUT FLOWER PRODUCTION IN THE PHILIPPINES

The Philippine Ornamental Horticulture Industry has come a long way since the 70's when cut flowers were considered its major component and the growers were hobbyists or plant enthusiasts. In the mid 80's the industry became a huge income source and potential for a foreign revenue earner. Interest in the commercialization of horticultural products grew to a point where the government recognized its importance and contribution to the local economy.

The number and size of farms had continued to increase adopting modern technologies where appropriate and in some cases foreign technology was sourced to boost production and quality. Promotion on the use of Ornamental Horticulture products increased people's awareness of the benefits that can be derived from the industry.

The floriculture industry in the Philippines is almost synonymous with the ornamental horticulture industry; the former connotes flowers; the latter, includes flowers and other ornamentals.

The country offers the following horticultural products for both domestic and external trade:

- Cut flowers - fresh flowers and flower buds that have been cut from the plant suitable for bouquets, wreaths, corsage and special flower arrangements.
- Foliage and other plant parts - fresh leaves and branches of trees, shrubs, bushes and other plants, grasses, mosses, lichens and the like suitable for ornamental purposes.
- Dried ornamentals - dried, dyed, bleached plant materials such as grasses, statice, eucalyptus.
- Other ornamentals - trees, shrubs, bushes, roots, cuttings and slips used indoor or for landscape purposes.

The phenomenal growth of the Ornamental Horticulture Industry in the last few years has led the Department of Agriculture to include the cut flower in its Key Commercial Crop Development Programme (KCCDP) for 1996 - 2000. The Department of Trade and Industry adopted the ornamental crops as its "Export Winner" and the Department of Science and Technology recognized it in its Science and Technology for National Development (STAND 2000). Young as it is, the industry has shown its vital role in the local economy, and slowly it is penetrating the export market and emerging as a true global competitor.

PRESENT SITUATION OF CUT FLOWER PRODUCTION

Cut flowers are grown throughout the Philippines but the industry was originally confined only to a few, small growers. In the last few years, an increased awareness and recognition of high return on investments, rapid population growth, higher standard of living, more hotels and restaurants, influx of tourists has led to more demanding and choosy clients. An increased demand triggered more production but despite the larger area devoted to cut flower, there is still a short fall in the supply. The demand for the domestic market is so big that the country has no option but to import some cut flowers, mainly chrysanthemum and orchids from other countries. This is strongly evident during Valentine's Day (Feb. 14), All Saint's Day (Nov. 1), School Graduation (March and April), May Flower Festivals and Yuletide Seasons (December). The major commercially grown cut flower varieties in the Philippines and their area of cultivation are shown in Tables 1 and 2.

In general, gladioli, aster, sampaguita, chrysanthemum and heliconias are grown in the open field. Depending on the cultural and light requirements of the cut flowers grown, greenhouses and net houses are usually constructed both in the highlands like Benguet and areas of low elevations. Semi-terete vandas are grown in the open, usually mounted on driftwoods; strap leaf vandas and dendrobiums on commercial basis are usually protected by 2 - 3 layers of fish nets and raised in pots/baskets. Anthurium production generally uses a more advanced technology - in greenhouse complete with computerized fertigation facilities. In some areas, net houses provide the necessary light requirements of the cut flowers. Roses are grown either in the open or in greenhouses, the same is true with the gerbera. Greenhouse structures protect the cut flowers from the intense heat of the sun and heavy rains. Cultural and management practices are also easier to control inside and can be scheduled regularly regardless of the weather outside.

The flowers of Jasminum sambac or sampaguita are usually used as leis or garlands usually with religious connotations. The plants are grown near houses or along roadsides for easy access and convenience on the part of the grower.

Flowering is controlled in chrysanthemum and aster production, usually in the form of supplementary lights. Sometimes growth regulators are applied in chrysanthemums if they are intended as potted flowering plants. Cold storage to break the dormancy of corms and cormels of gladiolus is available. Post harvest facilities for cut flowers are also available at the University of the Philippines, Los Banos. Big growers have their own storage facilities.

PRODUCTION OF PLANTING MATERIALS

Depending on the kind of cut flowers grown, planting materials are produced through cuttings, bulbs, tubers, corms, tissue culture, embryo culture, etc. For orchids, community pots, flasks/culture bottles, top cuts and overgrown seedlings are purchased either from abroad (Malaysia, Thailand, Hawaii) or from local nurseries. From the initial purchase, the cut flower grower can produce some of his planting materials for the farm.

Chrysanthemums are propagated mainly by cuttings, roses by budding, marcotting and cutting, anthurium by division or production of suckers, gladiolus through corms and cormels, heliconias by division. For big farms, the initial planting materials are imported, as in anthurium and chrysanthemum.

Table. Major Producers of Cut Flowers in the Philippines.

Cut Flower	Luzon	Visayas	Mindanao
Orchids	Laguna, Batangas	Cebu, Negros Occidental	Davao City, South Cotabato
Anthurium	Laguna, Benguet, Cavite, Batangas	Negros Occidental	Davao City Cagayan de Oro City, Bukidnon
Roses	Benguet, Cavite, Batangas, Tagaytay	Cebu, Iloilo, Negros Occidental	Davao City
Chrysanthemum	Laguna, Benguet	Cebu	Davao City
Gladiolus	Benguet, Laguna	Cebu	
Heliconias	Laguna	Negros Occidental	Zamboanga
Gerbera	Laguna, Batangas	Cebu	Davao City
Carnation	Benguet		
Aster	Laguna		
Shasta Dasies (Chrysanthemum maximum)	Benguet		
Zantedeschia aethiopica	Laguna, Batangas, Benguet	Cebu	
Jasminum sambac (Sampaguita)	Laguna, Pampangas		
Lilium (Oriental & Asiatic)	Laguna, Batangas, Benguet		Davao City
Statice (Limonium sinuatum)	Benguet		

Table. Varieties of Cut Flowers Commercially Grown in the Philippines.

Roses	Mums	Orchid	Anthurium	Gladiolus	Liliums	Heliconia
Red Success Mercedes Golden Times Texas White Liberty Jacaranda Raphaela Athena	Regan Series Fuji Series Taiwan Yellow Puto Puto	Jaq-Hawaii 'Uniwai Pearl' Jaquelyn Thomas Singapore White Sonia 'alba' Walter Oumae White Fairy Sonia Tuang Pink Waipahu Beauty Burana Jade Burana Fancy Chark Kuan Pink Chark Kuan Blue Chark Kuan Orange Chark Kuan Red Saleha Alsagoff Sumalee	Nitta Kaumana Chandler Midori Margaretha Fantasia Tropical Paradiso Lambada Merengue Leigh Mauricia	Friendship Spic & Span	Stargazer Cordelia	Parrot Parakeet Sase Jacquinii Bihai Caribbean Yellow Caribbean Red Kathy Southern Cross

Mist beds for rooting the cuttings and laboratories for embryo and tissue culture of orchids are available. Government agencies like universities and state colleges, regional offices of the Department of Agriculture and Department of Science and Technology usually offer services for embryo culture of hybrids produced by the grower for a small fee. In this manner, the grower can have his own planting materials. These agencies also sell their own produce to the public. Several growers specially the orchid breeders have their own laboratories and offer their hybrids for sale.

MARKETING

The key players in the marketing of the cut flowers in the Philippines are the breeders/growers, input suppliers, traders, cooperatives, transporters, brokers, exporters, importers, institutional buyers and walk-in consumers.

New and improved varieties of cut flowers form the backbone of the industry. Breeders, therefore, play an important role by developing the varieties and hybrids and help in the production of planting materials for the commercial growers. Input suppliers include the distributors and dealers of farm chemicals like fertilizers, pesticides, growth regulators and the like.

The traders can be individuals or groups of individuals who distribute cut flowers from the producers to the demand site. The traders could be assembler-wholesalers, assembler-wholesalers-retailers, commission agents, wholesalers, florists, retailers or vendors. Alternatively, members of cooperatives grow cut flowers and the organization itself may handle the selling and distribution of its produce to buyers. Transport of cut flowers is performed both by land and air *e.g.* planes, buses, vans, and jeepneys. Processing of papers of imported cut flowers and other supplies are performed by the brokers. Cut flowers are exported by corporations, organizations and individual entrepreneurs. Institutional buyers include hotels, restaurants, banks, offices, hospitals and churches. The quantity purchased by walk-in customers is minimal, depending on the occasion. Employees, students and housewives are the most common walk-in customers. There are no auction centres in the Philippines. There are only a few cooperatives that market their produce. Attempts have been made by some people to channel their cut flowers through their cooperatives but somehow, it is easier and more convenient to sell their produce individually.

Export of cut flowers is on a limited scale, with gladiolus topping the kind of cut flower exported. Almost all of it, which is about 46 per cent of the total export, goes to Korea. The other cut flowers exported to Japan, Brunei, Hongkong and Italy include anthuriums, liliums, and roses. More than half of the exported cut flowers go to Japan. A list of cut flowers imported in the Philippines. Chrysanthemums and carnations constitute the bulk of imported flowers from Australia, Holland and Malaysia. Almost half of the flowers are imported from Holland.

Table. Philippines Cut Flower Importation (1996).

Cut Flower	Country	Quantity (Kg)	% of Total Importation
Chrysanthemum	Australia	102	36.98
	Holland	8,389	
	Malaysia	13,566	
	Singapore	586	
Carnation	Australia	638	22.05
	Holland	7,691	
	Malaysia	5,097	
	New Zealand	80	
Pompom	Holland	176	7.29
	Malaysia	4,292	
Roses	Australia	55	6.81
	Holland	3,069	
	Malaysia	513	
	New Zealand	145	
	Thailand	60	
	USA	332	
Amaranthus	Australia	990	5.35
	Holland	2,226	
	New Zealand	60	
Alstroemeria	Australia	990	5.35
	Holland	2.226	
	New Zealand	60	
Lilium	Australia	301	4.19
	Holland	1,984	
	Malaysia	141	
	New Zealand	141	
Gerbera	Australia	78	3.79
	Holland	1,976	
	Malaysia	257	
	New Zealand	15	
Anthurium	Australia	428	3.58
	Holland	1,545	
	Malaysia	202	
	New Zealand	10	
Tulip	Australia	57	2.95
	Holland	1,629	
	New Zealand	71	
	USA	50	
Misty Blue	Australia	72	1.60
	Holland	61	
	Malaysia	852	
TOTAL			61,224

Table. Cut Flowers Exported, 1996.

Cut Flower	Country	% Share
Gladiolus	Korea	46.10
Anthurium	Brunei, Japan	43.49
Liliums	Japan	8.93
Roses	Hongkong, Japan	1.33
Orchids	Italy, Japan	0.41

Table. Export Markets of Philippines Cut Flowers for 1996.

Country	Cut Flower	Quantity (Kg)	% Share
Japan	Anthurium	9,358.62	53.48
	Liliums	1,933.92	
	Roses	279.50	
	Orchid	6.00	
Korea	Gladiolus	10,000.00	46.19
Brunei	Anthurium	55.62	0.25
Hongkong	Roses	10.00	0.046
Italy	Orchids	3.00	0.013

Table. Suppliers of Cut Flowers to the Philippines (1996).

Country	Cut Flower	%Share
Holland	Tulip, Roses, Protea, Pompom, Peacock, Misty blue, Limonium, Lilium, Iris, Holland, Gypsophila, Gladiolus, Gerbera, Freesia, Banksia, Alstroemeria, Aster, Anthurium, Amaranthus	50.83
Malaysia	Roses, Pompom, Phoenix, Peacock, Misty Blue, Limonium, Lilium, Gerbera, Freesia, Chrysanthemum, Carnation, Baby's Breath, Aster, Anthurium	39.75
Australia	Amaranthus, Anthurium, Alstroemeria, Baby's Breath, Banksia, Carnation, Chrysanthemum, Freesia, Gerbera, Iris, Lilium, Misty Blue, Protea, Roses, Tulip	6.84
New Zealand	Amaranthus, Anthurium, Alstroemeria, Banksia, Carnation, Freesia, Gerbera, Lilium, Roses, Tulip	.98
Singapore	Chrysanthemum	.89
USA	Roses, Tulip	.58
Thailand	Anthurium, Roses	.10

POTENTIAL FOR CUT FLOWER PRODUCTION DEVELOPMENT

The country has its strengths for cut flower production. It has an ideal climate for year round production with the capability to grow both tropical and subtropical varieties.

Considering the wide availability of land waiting for tillage and production of agricultural crops, it would not be difficult to produce cut flowers since suitable production techniques are already available. The richness and diversity of Philippine flora allows the development of varieties of cut flowers with endemic species as parents. A very important factor that contributes to the cut flower production in the Philippines is the existence of a very active horticulture association, of which commercial growing of cut flowers is one of the main objectives.

The Federation of Cut flower and Ornamental Plant Growers, a non-stock, non-profit private organization was formed in 1990 with the support of the Department of Agriculture to assist the government develop the cut flower industry. Its main thrust is service in the different fields where technical manpower support is lacking. The services are on consultation, skills and training, involvement in the marketing of the products of the growers, establish linkages with non-government organizations like USAID, JICA and Dutch-Flamingo International (FMD). The Federation is involved in advocating new policies for the improvement of the industry and its members.

There is still a substantial shortage of supply of cut flowers in the Philippines. Importation of cut flowers from other countries is quite substantial . Hence, by stepping up the cut flower production within the country, a substantial amount of foreign exchange can be saved.

CONSTRAINTS IN CUT FLOWER PRODUCTION DEVELOPMENT

The high cost of structures like greenhouse, irrigation and postharvest facilities are some of the major constraints to the producers of cut flowers. The availability of quality planting materials most of which come from abroad is sporadic. For new varieties/types of cut flowers like lilium, gerbera and calla, production technology is still insufficient. With new production technology comes the introduction of new agricultural chemicals which are not readily available locally and work out very expensive when imported. This is coupled with high import duties on other inputs too.

Access to credit sources is also another major problem confronting the growers. The interest rates on loans offered by institutions are high. Growers of roses, chrysanthemums, gladiolus, orchids and anthuriums are in need of financial support during the initial stages of production. Collateral requirements demanded by lending agencies are rarely met and the grower faces the dilemma of whether to continue cut flower growing or not.

On the part of the exporters, there is a shortage of cut flowers volume wise and quality wise. The small growers can not compete with big growers. It will be easier to meet the quality, but the volume needs more attention.

The cut flower industry is still in its juvenile stage and not much information on the feasibility and prospect of the business is available. Hence, investors are slow in getting into the business since lending agencies or banks always demand feasibility studies. Big growers are quite hesitant to provide information because of inherent competition.

CUT FLOWER PRODUCTION IN SRI LANKA

Cultivation of flowers for various religious and cultural festivals has existed in Sri Lanka for ages. It was only after the British rule that floriculture really came into practice not as an industry but mainly as a hobby for pleasure. Later, with many new introductions of tropical and sub tropical plants the trend was gradually passed down to other levels of society. Floriculture in Sri Lanka started as an industry in 1970. It has grown substantially during the last few years to become one of Sri Lanka's major foreign exchange generating ventures.

PRESENT SITUATION OF CUT FLOWER PRODUCTION

Western, North Western and Central Provinces in Sri Lanka are the major areas where cut flowers are grown commercially. Cut flowers grown in the country can be divided into two main categories based on their temperature requirements *i.e.* Temperate Cut Flowers and Tropical Cut Flowers.

Temperate Cut Flowers: Temperate cut flowers include carnation, rose, statice, gypsophyla, alstroemeria, chrysanthemum, lilies and irises. Among them, carnations and roses are grown mainly for export, in the highlands of the Central Province of Sri Lanka. Carnations are produced entirely from imported planting materials and are graded according to internationally accepted specifications for export. The other species are commercially grown mainly for the local market.

American and Mediterranean carnation cultivars are quite famous in the world market. Pink, white, red, yellow and salmon colours are much popular. Novelties such as striped and frosted types are also becoming increasingly popular. American cultivars grown are silvery pink, karina, barbara, red barron, elsy, royalette, bagatalle, bianca and adelfie. Mediterranean cultivars grown are nora barlo, shainah, lena, castellaro, scania, tanga, roma, pallas and charmeur. Carnations are grown in poly tunnels, covering more than 10 hectares, under fully protected environments.

Roses are second to carnations and production is limited to about 40,000 blossoms per annum. This is mainly due to the highly expert conditions required for cultivation and production. Approximately 2-3 hectares of area have so far been used for cultivation of roses. Roses for the export market are grown under controlled environments in poly tunnels. Popular colours are highly variable. They are generally required in a mix and an acceptable mix would consist of red-50 per cent, Pink-30 per cent, yellow-10 per cent and others 10 per cent. Roses are quite popular in the local market also. Most of them are supplied by small scale growers.

Tropical Cut Flowers: Anthuriums and Orchids are the most popular tropical cut flowers which are being grown commercially for exports as well as for the local market. Anthuriums can be grown at elevations up to 1500 metres above sea level, with texture and the flossiness remarkably enhanced with increase in elevation. Annual production of anthuriums is around 3 million flowers, the majority of which are sold at the local market. The exports of Anthuriums at present are not very significant. A land area of approximately 10 hectares is under Anthurium cultivation at present and the industry is expanding steadily at village level. Almost all plantations are either under poly tunnels or structures with shade netting. However, locally available materials are also used under certain conditions to provide required shade levels.

The standard trade types in the world market are the "Avo" lines *i.e.* Avo Nette, Avo Ingrid, Avo Anneke. The improved standard types *e.g.* Germa, Cuba, Fuego, Favouriet etc. are also popular. The Royal Botanic Gardens, Peradeniya, being the pioneer institution for the development of cut flower industry in Sri Lanka has produced a few promising Anthurium varieties with export potential. Some of these varieties are being closely studied and mass propagation has just been started. A few of the slected varieties are: RBG - Green Tip, RBG - Soft Sheen, RBG - kandy queen, RBG - Gardens Pride, RBG - Royal Flag, RBG - Lak Isuru, RBG - Wild Beauty, RBG - Krishnas Red.

Tropical Orchids can be grown under warm humid conditions up to 500 metres above sea level. Climatic conditions in the Western Province are quite favourable for cultivation of Dendrobiums, Vandas and Phalaenopsis types which are quite popular in the local and foreign markets. Approximately 3-4 hectares of land are under Orchid cultivation at present and the industry is gradually developing to cater to export markets. Almost all cultivations are under shade netting and the majority of growers have developed mist irrigation techniques. 'Madam Pompadour' (Pink and White), and 'Rena Vapahoo' (Pink and White) hybrids are some of the popular hybrid Dendrobiums in the market. Hybrids of Arachnis, Oncidiums, Phalaenopsis and Dendrobiums are being exported in smaller quantities and the income earned is not very significant.

Gerberas are becoming popular among growers due to the availability of a wide range of long lasting cultivars produced by many modern breeding methods and tissue culture. These cultivars can be broadly categorised into three classes, namely Singles (such as 'Fleur' and 'Apple Blossom'), Doubles (such as 'Marleen' and 'Hildegard'), and Black Centres (such as 'Fabio' and 'Rosseta'). Pink, salmon, orange, red and yellow are the popular colours in the market. Trials have already been started under controlled environments by private entrepreneurs to grow gerberas for the export market. About 2 hectares of land is so far being used for gerberas.

GROWER CATEGORIES AND SPECIALITIES

Generally, large scale commercial establishments also have their cultivations under green house conditions, poly tunnels or netting. The medium and small scale growers either have their cultivations under natural shading (under trees) or use locally available materials such as coir fibre mats, cajans (dried and woven coconut palm leaves) or ropes to provide shade requirements of a particular crop. Sri Lanka's floriculture industry consists of three categories of producers or growers a) large commercial ventures for export; b) middle level growers catering to the local market; and, c) village level producers who may sell their products to either of the two categories mentioned above. The total land area under floriculture is around 500 hectares at present and the majority of lands are in the Western Province. There are 10 ha under carnations, 3 ha under roses, 2 ha under gerberas, 10 ha under anthuriums, 3 ha under orchids and 472 ha under foliage plants. These figures clearly show that the floriculture trend so far in Sri Lanka has been in favour of foliage plants. The majority of large scale commercial growers produce plants in collaboration with foreign partners. Technology is shared by these partners and highly advanced methods of production are followed. However, most of these ventures are involved in the production of ornamental foliage plants and very few are involved in the production of cut flowers. Middle level and village level growers usually go in for low cost cultivation with minimum advanced techniques, sticking to conventional methods. The majority of these growers are the ones who produce cut flowers. Therefore, most of the cut flowers produced are used to satisfy the local needs and only the surplus is exported occasionally in small amounts. The exception is the production of carnations which are grown mainly for the export market. However, export oriented cut flower ventures are on the increase.

PRODUCTION OF PLANTING MATERIAL

Flower seeds of many flowering annuals such as asters, petunias, impatiens, phlox and verbenas are produced mainly for export under fully controlled green house conditions. A small part of this production is released to the local market at regular intervals. The bulk of floriculture planting materials exported are in the form of stem cuttings rooted or unrooted. A very small amount of bulbs, corms and tubers are exported to specific customers abroad.

Large tissue culture units are few in number. However, at present, six commercial tissue culture units are operational. One of these laboratories specialises in the production of cut flower anthurium plants through tissue culture. This venture is of recent origin and produces large quantities of anthurium plants now popular in the world market. Four of these laboratories are attached to large commercial, export oriented nurseries involved mainly in ornamental foliage plants and to a lesser extent cut flowers. The sixth of these laboratories is exclusively for the production of orchids for export. There are a few middle level orchid growers who own mini-laboratories for the production of orchid seedlings and other ornamental plants for village growers.

MARKETING

Thus far the cut flower market in Sri Lanka has not been able to create Auction Centres as many other countries have done. Retail outlets scattered through the production areas are the popular centres where cut flowers are sold. There are few growers who have created cooperative systems to sell their products.

Exporting of cut flowers is done by a few societies which have a selected group of partners and farmers. In many instances the agents are sent to villages to collect flowers from farmers directly. Exports of floriculture products from Sri Lanka. Main export markets are Europe (72 per cent), and Far East and Middle East (28 per cent).

Table. Export of Floriculture Products from Sri Lanka (1990-1995).

Product	Value (Million Rupees)					
	1990	1991	1992	1993	1994	1995
Bulbs, Corms, Tubers	0.5	0.71	21.2	21.6	10.03	6.1
Live Plants	99.5	123.1	132.4	120.3	162.4	179.0
Cut Flowers	31.0	29.7	21.7	34.5	45.3	54.8
Cut Foliage	60.3	77.74	106.9	111.0	189.1	190.2
Total Rs. Million	191.3	231.25	282.2	287.4	406.83	430.1

POTENTIAL FOR CUT FLOWER PRODUCTION DEVELOPMENT

Sri Lanka's stable and varied tropical climatic conditions and the geographic terrains from sea level up to 2200 metres of humid mounts have created magnificent macro and micro environments to house many thousands of local as well as sub-tropical and temperate plants. The rich native flora contains many potential ornamental plants which could be developed to satisfy the demands of the flourishing industry.

Sri Lanka has a favourable location to serve different markets in the world. Availability of land and the high literacy rate of the average person would be an added benefit to those who wish to invest in the industry. In addition, the tax benefits and BOI incentives granted by the government would help bring in more investors to the country and facilitate further development of the existing industry. Furthermore, the new policy framework prepared by the Ministry of Agriculture and Lands has clearly identified the need to initiate a Floriculture Research and Development Programme. This programme will cover many areas such as agronomy, pathology, entomology, mutation, breeding, post harvest and mass propagation to support the industry.

CONSTRAINTS IN CUT FLOWER PRODUCTION DEVELOPMENT

The following are the major constraints being faced by the majority of growers:

a. *Inadequacy and high cost of air cargo:* Air Lanka, the national carrier has always given priority to perishable cargo. However, the available capacity is not sufficient.
b. *Lack of facilities for research and development:* So far only the Royal Botanic Garden, Peradeniya has been involved in providing assistance by and large to the middle and village level growers, but this is negligible when compared to the ever increasing demands of the industry.

c. *Lack of trained personnel:* The floriculture industry requires trained personnel at each level of production. Education programmes from schools up to University level and training institutes to conduct courses on high-tech practical skills in floriculture are essential for the development of the industry.
d. *Big initial investment on farms:* Duty free facilities for import of vital items not produced in Sri Lanka such as shade nets, uv stabilised polythene, irrigation and fertilising systems etc. would help to promote the industry.
e. *Lack of improved systems of marketing:* Lack of proper or organised systems for marketing and inadequacy of current international market information on prices, trends, volumes and data on competitive countries etc, severely affect the development of the industry.
f. *The uneconomical size of floriculture industry:* Difficulty in acquiring suitable land and lack of infrastructure facilities also adversely affect the industry.
g. *Lack of information on pesticides:* Current information on pesticides and their acceptance in various countries is essential.
h. *Phytosanitary clearance:* Phytosanitary inspection just prior to shipment are inconvenient and expensive, besides leaving no time to rectify any problems.

CONCLUSIONS

Floriculture has a great potential in Sri Lanka even though the existing market share is less than 0.2 per cent of the world market at present. Interest in the cut flower industry is encouraging and many investors, both local and foreign, have started their nurseries to cater to the expanding market.

A comparison of the world trend and the production figures of Sri Lanka clearly show that we are not in line with the world trend.

However, this situation will soon change with the new policy changes taking place in the country.

The Ministry of Agriculture and Lands, Export Development Board and various Floricultural Associations of the country have got together to discuss various issues in this regard and progress so far is very encouraging. Steps should, however, be taken to encourage floriculture research to cater to the special needs of the existing industry and to facilitate long term research programmes to explore the rich flora of the country to develop novelties to satisfy the growing market.

The Government should take immediate steps to alleviate constraints mentioned above and to create an Institute or a Centre for Floriculture Research and Development. This could be a semi-governmental Centre or an Institute partially supported from the growers annual income.

CUT FLOWER PRODUCTION IN THAILAND

Administratively, Thailand is divided into 76 provinces and 5 regions (North, Northeast, Centre, East and South). The country has abundant water supply and suitable agro-climatic conditions for the production of various kinds of orchids and other cut flowers. Depending on the location, tropical, subtropical and temperate cut flower species can be grown. Bangkok and the nearby provinces are good for orchids and other tropical flowers, whereas the subtropical and temperate cut flower species are best grown in the north and northeastern parts of the country, especially in the winter season.

The country was the world's fourth largest exporter of cut flowers in 1993 and 1994. Exports consisted almost exclusively of a wide range of cut orchids and amounted to 748.65 million baht (US$ 65.37 million) in 1993 and to 782.45 million baht (US$ 68.20 million) in 1994. The volume of cut orchids exported in 1994 was 11,897 tons.

Table. World Cut Flower Exports Ranked by Source Country (US $ million) (1992-1994).

Rank	1992		1993		1994	
	Country	Value	Country	Value	Country	Value
1.	Netherlands	2,153.56	Netherlands	1,456.24	Netherlands	1,586.40
2.	Colombia	395.64	Colombia	415.61	Colombia	431.71
3.	Israel	146.12	Israel	115.26	Israel	134.15
4.	Italy	111.28	Thailand	65.37	Thailand	68.20
5.	Thailand	67.58	Kenya	60.57	Kenya	67.57
6.	Kenya	61.48	Italy	48.39	Ecuador	52.88
7.	Others	388.17	Others	318.94	Others	400.95
	Total	3,323.82	Total	2,480.37	Total	2,741.86

Remark: 1 US$ = 25 baht (at 1995 rate of exchange)

Table. Quantity and Value of Cut Flower Exports from Thailand (1988-1994) Quantity: Ton; Value: × 1,000 Baht.

Year	Cut Orchids		Other Cut Flowers		Total	
	Quantity	Value	Quantity	Value	Quantity	Value
1988	9,532	515,809	14	1,107	9,546	516,916
1989	10,752	506,898	38	1,210	10,790	508,108
1990	11,678	552,093	38	1,336	11,716	553,429
1991	12,399	662,383	35	1,530	12,434	663,913
1992	11,142	701,328	9	478	11,151	701,806
1993	12,375	748,648	9	1,061	12,384	749,709
1994	11,897	782,446	20	1,268	11,927	783,724
% Change	+3.26	+8.44	-10.96	-3.11	+3.27	+8.42

In 1994, Thailand exported orchid cut flowers to Japan, Italy, U.S.A., Germany, Taiwan and the Netherlands. During the same year the country also exported garlands made from orchids to U.S.A., Germany and Taiwan. Besides cut orchids, Thailand exported other cut flowers which were mainly tropical flowers, though the volume was very small. The flowers exported included Jatrophia, Lotus, Ananas, Heliconia, Jasmine, Rose and Marigold.

Although Thailand exported a lot of orchids, it also imported small quantities of cut orchids from Netherlands, India, Malaysia and Singapore, in addition to other cut flowers, mainly temperate, which were imported in larger volumes. The total value of imports in 1994 was 63.58 million baht.

PRESENT SITUATION OF CUT FLOWER PRODUCTION

Categories of Cut Flowers

The cut flower production can be divided into two categories *i.e.* cut orchids and other cut flowers.

Cut Orchids

The production area of orchids was 13,000 rai (6.25 rai = 1 ha) in 1988 and increased to 14,412 rai in 1994. The production was 18,750 tons in 1988 and increased to 25,900 tons in 1994. Main production areas were Samut Sakhon, Nakhon Pathom, Bangkok, Ratchaburi, Pathum Thani and Ayuthaya.

Table. Production Area of Cut Orchids and Other Cut Flowers in Thailand (1988-1994).

Year	Cut Orchids			Other Cut Flowers			Total		
	Total Area (Rai)	**Total Prod. (Ton)**	**Prod./ Rai (Kg.)**	**Total Area (Rai)**	**Total Prod. (Ton)**	**Prod./Rai (Kg.)**	**Total Area (Rai)**	**Total Prod. (Ton)**	**Prod./Rai (Kg.)**
1988	13,000	18,750	1,442	21,045	27,224	1,294	34,045	45,974	1,350
1989	13,100	17,950	1,370	20,271	28,162	1,389	33,371	46,112	1,382
1990	13,100	19,650	1,500	20,071	27,756	1,383	33,171	47,406	1,429
1991	13,200	22,500	1,704	23,700	30,501	1,287	36,900	53,001	1,436
1992	12,534	21,625	1,725	22,316	24,894	1,116	34,850	46,519	1,335
1993	12,614	21,750	1,724	25,700	28,500	1,109	38,314	50,250	1,312
1994	14,412	25,900	1,797	26,981	31,450	1,166	41,393	57,350	1,385
Change	+0.68%	+5.33%	+4.60%	+4.85%	+1.25%	-3.43%	+3.31%	+2.96%	-0.34%

Remarks:

Cut orchids 40 stems/kilogram

Other cut flowers 35 flowers or stems/kilogram

Jasmine 650 kilogram/rai

Orchid cultivars grown for cut flowers are:

Dendrobium:	Anna, Blue, Bangkok Land, Big White, Candy, Candy Kiss, Casablanca, Caesar, Channel, Chidchom, Intuwong, Juliana, Madame Pompadour, Marco Polo, Mary Mak, Missteen, Nina, Royal Pink, Sabin, Sakura, Siam Ruby, Sonia, Walter Oumae, Yoko, Waipahu
Aranda:	Christine, Choo Lai Keun, Nora Blue
Aranthera:	James Storei
Renanthera:	Azimah, Okahara
Mokara:	Jark Kuan
Oncidium:	Golden Shower, Gold Star
Vanda:	Alice Blue

The production cost of Dendrobium cut flowers was 178,741 baht / rai, or 2.27 baht per stem. The income of the farmer was 194,242 baht/rai and the profit 17,501 baht/rai.

Other Cut Flowers

Jasmine: The area had increased from 3,600 rai in 1988 to 5,326 rai by 1994. The annual production was 6,718 tons in 1988 and increased to 9,447 tons in 1994. Main production areas were Nakhon Pathom, Nakhon Sawan, Bangkok, Samut Sakhon. The production cost per rai was 75,116 baht or 40.26 baht per kilogram. Average income was 104,365 baht per rai. The profit was 29,249 baht per rai. Rose: During the years 1988-1993 the production area decreased from 3,500 rai to 2,231 rai. In 1994 it increased to 3,911 rai. The annual production had decreased from 11,257 tons in 1988 to 9,457 tons by 1994. Main production areas were Nakhon Pathom (1,377 rai), Samut Sakhon (905 rai), Nonthaburi (610 rai), Bangkok (235 rai), Ratchaburi (150 rai), and Chiang Mai (106 rai). The average cost of production was 87,883 baht per rai, or 0.76 baht per stem. The profit was 98,050 baht per rai. Gerbera: The production area of double-strain gerbera had decreased rapidly from 2,000 rai in 1988 to 368 rai by 1994, mainly because the growers changed from the old-fashion double-strain gerbera to European strain with longer vase life. The European strain which is single is more suitable to grow on the highland and is in demand in the local market. Major production areas were Ratchaburi (100 rai), Nonthaburi (64 rai), Phattalung (45 rai), Ranong (25 rai) and Chiang Mai (23 rai). Cost of production was 81,347 baht per rai or 1.09 baht per stem. Average income per rai was 154,859 baht, and the profit was 73,511 baht per rai. Chrysanthemum: The total area was 580 rai in 1988 and had increased to 998 rai by 1994. The annual production was 1,204 tons in 1988, and had increased to 2,886 tons by 1994. Main production areas were non-thaburi (432 rai), Chiang Mai (194 rai), Yala (53 rai), Nakhon Phanom (42 rai), Phat Thalung (40 rai). The average cost of production was 59,940 baht per rai or 1.64 baht/stem. The income was 110,634 baht per rai. The profit was 50,694 baht per rai.

Aster Peacock: The production area of 200 rai in 1988 had increased to 407 rai by 1994, and the annual production from 256 tons in 1988 to 529 tons by 1994. Main production areas were Chiang Mai (213 rai), Nong Khai (57 rai), Nonthaburi (35 rai) and Songkhla (31 rai). Production cost per rai was 35,921 baht or 41.21 baht per kilogram. Average income was 48,633 baht per rai. The profit was 12,712 baht per rai.

Production Area

The total area of cut flowers in the year 1994 was 39,893 rai. The largest growing area was Bangkok with 8,853 rai, followed by Samut Sakhon, Nakhon Pathom and Nonthaburi with 7,097 rai, 6,780 rai and 5,735 rai respectively.

Table. Cut Flower Production Area by Province in Thailand (1994).

Rank	Province	Area (Rai)	Percent
1	Bangkok	8,853	22.19
2	Samutsakorn	7,097	17.79
3	Nakhonpathom	6,780	17.00
4	Nonthaburi	5,735	14.38
5	Phayao	1,397	3.50
6	Nakhonsawan	1,300	3.26
7	Ratchaburi	1,033	2.59
8	Lampang	982	2.46
9	Chiangmai	869	2.18
10	Suphanburi	743	1.86
11	Kanchanaburi	385	0.96
12	Kamphaengphet	282	0.71
13	Ranong	277	0.69
14	Songkhla	268	0.67
15	Udonthani	244	0.61
16	Other Provinces	3,648	9.14
	Total Area	39,893	100

gives the species-wise area for cut flowers grown in 1994. The major kinds were orchids - 14,412 rai, jasmine - 5,326 rai, lotus - 4,371 rai, marigold - 4,072 rai, rose - 3,911 rai, giant milk weed - 2,054 rai, champaca - 1,725 rai and chrysanthemum - 998 rai; the other cut flowers included golden rod, aster peacock, gerbera, torch ginger, gladiolus, heliconia, globe amaranth, anthurium, curcuma, tuberose, red ginger and jatrophia. The main production area of orchids and other tropical cut flowers were Bangkok and the nearby provinces; while the temperate cut flowers namely lily, freesia, tulip, calla lily, alstroemeria, bird of paradise, statice, gypsophilla and carnation were grown in the Northern part (Chiang Mai and Chiang Rai) and the Northeastern Part (Nakhon Ratchasima, Ubon Ratchani, Udon Thani, Kohon Kaen, Nong Khai and Nakhon Phanom).

Cultivation Conditions

Cut Flowers Grown in the Open Field: Cut flowers which are grown in the open field are: aster peacock, champaca, curcuma, giant milk weed, gladiolus, globe amaranth, heliconia, jasmine, lotus (in water), marigold, tuberose.

Cut Flowers Grown under Protected Conditions: Cut flowers which are grown under protected conditions are:

Shading house (saran) 50% shading:	Orchids
60% shading:	Red ginger, Torch ginger
70% shading:	Anthurium
Rain-proof house (Vinyl house):	Carnation, Chrysanthemum, Gerbera, Lily, Rose
Insect-prevent house (Net house):	Chrysanthemum

Table. Cut Flower Production Area by Kind of Flowers Grown in Thailand (1994).

Rank	Flowers	Area (Rai)	Percent
1	Orchid	14,412	36.16
2	Jasmine	5,326	13.36
3	Lotus	4,371	10.97
4	Marigold	4,072	10.22
5	Rose	3,911	9.81
6	Giant Milk Weed	2,054	5.15
7	Champaca	1,725	4.33
8	Chrysanthemum	998	2.50
9	Golden Rod	644	1.61
10	Aster (Peacock)	407	1.02
11	Gerbera	368	0.92
12	Torch Ginger	191	0.48
13	Gladiolus	104	0.26
14	Heliconia	74	0.19
15	Globe Amaranth	63	0.16
16	Anthurium	41	0.10
17	Curcuma	32	0.08
18	Tuberose	24	0.06
19	Red Ginger	10	0.02
20	Others	1,025	2.57
	Total Area	39,893 rai	

Remark: 6.25 rai = 1 hectare

Advances in Production Technology

Breeding and Selection: Many Universities, the Ministry of Agriculture, the Royal Projects and many companies in the private sector have done good work in improving, introducing and selecting good cultivars of cut flowers for the market. The methods employed for different crops included:

Hybridization:	Orchid, Gerbera, Lily, Anthurium, Curcuma
Somaclonal variation:	Orchid
Induced mutation:	Orchid, Carnation, Chrysanthemum
Induced polyploidy:	Curcuma and Lily
Protoplast fusion and genetic engineering:	Orchid

Micropropagation: Mass rapid clonal propagation using tissue culture techniques has been employed for commercial multiplication of plant material for the growers in orchid, gerbera, carnation, chrysanthemum, anthurium, curcuma, red ginger, torch ginger, lily, calla lily. There are around 50 commercial tissue culture units in Thailand.

Irrigation Technology: Advanced irrigation methods like sprinkler (for orchid and anthurium) and drip irrigation (for rose, gerbera, carnation, chrysanthemum, lily) are used for automatic watering.

Production Cycle Management: Technology for off-season and year round production for various flower crops were standardised using photoperiodic control (chrysanthemum, aster peacock, curcuma), temperature control (lily, gladiolus) and pinching technique (orchid).

Post Harvest Technology: Research efforts were directed to standardization of post harvest technologies related to improving vase life of cut flowers through using appropriate holding solutions or controlling atmosphere/temperature in cold storage.

PRODUCTION OF PLANTING MATERIAL

The requirement of planting material for cut flower production was mainly met through introduction of seed (marigold, statice, bird of paradise), bulbs (lily, gladiolus, calla lily), cuttings (carnation, chrysanthemum, jasmine), divisions (gerbera, aster peacock, red ginger, heliconia, torch ginger, alstroemeria, tuberose) and budding and grafting (rose). In addition, as mentioned earlier, micropropagation (tissue culture) was used for mass production of planting material for various species.

MARKETING

Most cut flowers are for domestic consumption, except orchids 50 per cent of which are exported. During the period 1988-1994, the demand for domestic consumption of orchids and other cut flowers increased from 33,696 tons to 46,330 tons.

Table. Demand and Supply of Cut Flowers in Thailand (1988-1994).

Year	Demand		Total	Supply	
	Domestic Prod.	Imports		Domestic Cons.	Exports
1988	45,974	268	46,242	33,696	9,546
1989	46,112	319	46,431	35,641	10,790
1990	47,406	373	47,779	36,063	11,716
1991	53,001	364	53,365	40,931	12,434
1992	46,519	504	47,023	35,872	11,151
1993	50,250	473	50,723	38,339	12,384
1994	57,350	897	58,247	46,330	11,917
% Change	+2.96	+18.33	+3.09	+3.05	+3.24

Remarks: Domestic Consumption = Domestic Production + Imports - Exports

For local consumption, there is demand for high quality flowers by hotels, restaurants and florist shops. The flowers must be attractive, large, colourful, with long vase life. The demand is all year round. Other consumer groups use flowers for religious purpose, where the quality of flowers does not have to be so high, and cheap price is more important. The cut flower demand is changing according to the occasions in Thai society.

The price of cut flowers depends on grade and quality, the season and demand during special occasions, like Buddhism days and Graduation days. During 1994-95 the average price for major cut flowers was: for orchids 2.49 baht/stem; rose 1.61 baht/stem; chrysanthemum 3.02 baht/stem; gerbera 2.07 baht/flower; aster peacock 1.61 baht/stem; jasmine flower 55.93 baht/kilogram.

CONSTRAINTS IN CUT FLOWER PRODUCTION DEVELOPMENT

Although Thailand has the fourth rank in world export of cut flowers, there are several constraints to be considered for future development of this industry.

These are as follows:

- The multiplication rate of the selected cultivars by conventional methods is slow; therefore, tissue culture micropropagation is used.
- High initial investment, especially the cultivars that have to be imported or grown under controlled greenhouse conditions.
- Increasing labour cost.
- Short vase life.
- Production depends on the season, and not always on the market demand. Sometimes there are not enough flowers, and sometimes there are too many flowers, which results in unstable prices.
- Improper application of chemicals for controlling diseases and pests.
- Post-harvest technology and packaging suitable for Thailand are not available.
- No cool storage at the airport.
- Insufficient air cargo space during the peak season for export.
- Lack of efficient systems in communication and information, resulting in difficulty to plan production according to the market demands.
- The strict phytosanitary requirements enforced in the importing countries.

POTENTIAL FOR CUT FLOWER PRODUCTION DEVELOPMENT

Thailand has the capability to develop a sound cut flower industry for export due to favourable climatic and infrastructural conditions, cheap land, low labour cost, relatively low capital investment and high value addition. Future potential for the cut flower industry in Thailand is very good, but it is necessary to ensure that the flowers are of high quality, there is a reliable and continuous supply of flowers, sufficient quantities of selected cultivars and competitive prices.

The Eighth Economic Development Plan for cut flower development in Thailand in the next five years (1997-2001) indicates that the production area should increase by 2.67 per cent, the production per rai by 5.63 per cent and the export value by 4 per cent. The production area of cut flowers was 45,000 rai in 1997 and is expected to increase to 50,000 rai in 2001. The total annual production volume is 62,550 tons, and is expected to increase to 86,850 tons.

CONCLUSIONS

At present, Thailand has the fourth rank in the world cut flower exports market with orchid as the main crop. Orchid cut flowers form only a relatively small fraction of the world cut flower market. The top-seller cut flowers in the world market like carnation, chrysanthemum, rose, gypsophyla, gladiolus, asiatic lily, tulip, iris, oriental lily, daffodil and cape flower, can also be grown in Thailand. Research and development efforts have to be strongly supported for the improvement of the cut flower industry in the country.

CUT FLOWER PRODUCTION IN VIETNAM

Vietnam is located in South East Asia between latitudes 80 and 230 North. The total land area is 33,099,093 hectares (330,990 km^2). Its population is approximately 76 million (1996). The country can be divided on the basis of its topography, soil pattern and climate into nine agro-ecological zones.

The country has a tropical monsoon climate with a pronounced dry winter season. The Northern delta has a cool winter from November to April with average temperature around 16°C, and frequent light drizzle from February onwards. Summer in Northern provinces is very hot, rainy and subject to typhoon. The Southern delta has a more equable monsoon with an average temperature of 25 - 30°C. Rainfall in most areas varies from 1,600 to 2,500 mm per year. Soils are formed from parent rocks in the NE and NW, alluvium in the deltas, chronic buvisols in the coastal lowlands and fertile soils (basalt) in the central highlands. The soil and climatic conditions in Vietnam are very suitable for flower cultivation round the year, though presently it occupies a very small share in agriculture.

Table. Agricultural Land Use in Vietnam (1993).

Item	Area (ha)	Share (%)
I. Annual Species	5,523,899	75.00
a. Rice	4,252,107	57.80
b. Other Cereals	1,124,542	15.30
c. Vegetables	15,158	0.20
d. Flowers	1,585	0.02
e. Other Annual Species	130,081	1.73
II. Perennial Species	1,247,161	16.90
a. Industrial Crops	525,342	7.10
b. Fruit Trees	187,309	2.50
c. Other Perennial Species	532,767	7.20
d. Nurseries	1,743	0.02
III. Grazing Land	304,274	4.10
IV. Water Surface for Agriculture Use	273,115	4.00
Total	7,348,449	100.00

The flower production area covers only 0.02 per cent of the total cultivated area, even though flower production is a traditional activity in Vietnam. Flowers are mostly cultivated in urban areas of Haiphong, Hanoi, Ho Chi Minh, Dalat and provincial towns.

In the past decades, flower and ornamental production has been the business of small holders and could not be established into a viable industry, largely owing to a limited market. To make it a profitable industry, flower producers must be oriented to accessions of festivals, while the profit from flower production should attain the level of other crops.

Since 1986, with the economic renovation process, the situation of the country has changed as the life of the majority of the population has improved significantly.

This is reflected on flower production through a significant increase in the area under flower cultivation, with new villages/regions becoming involved in flower production; also with product diversification with orchids, chrysanthemums, lilies etc. being introduced for commercial production; furthermore with research on flowers being started for variety improvement and standardisation of improved production technologies.

The flower industry in Vietnam is about to be established and form a new economic area involving thousands of farmers. This process will be accompanied by the economic growth of the country and raised living standards of the people. Foreign investments can contribute in this process and Vietnam flower producers could have a share in the international market.

PRESENT SITUATION OF FLOWER PRODUCTION

Production Area

The area of cut flower production in Vietnam at present is rather small. It is estimated that only 1,585 hectares are under cut flower production, which cover about 0.02 per cent of the total agricultural land. The major production areas are concentrated in big cities such as Hanoi, Ho Chi Minh City, Hai Phong and Dalat. The area under cut flowers has increased significantly during recent years and is likely to increase further in coming years. However, still big specialised cut flower farms do not exist. Cut flowers are mainly produced by small holders in their gardens.

To meet the increased demand of flowers, the area under cut flower production needs to be increased through involvement of new producers, and the existing producers must increase their production area by renting or buying new land, formerly under other crops. It is expected that the production area will increase substantially if Vietnamese cut flower producers can enter the export market. The explosion of population, urbanization and industrialization also increases the demand and market for cut flowers.

Table. Area Under Flower Cultivation in Vietnam.

Province	Area (ha)
Hanoi	500
Haiphong	320
Ho Chi Minh	200
Dalat	75
Others	490
Total	1,585

Flowers are cultivated in Ngocha, Quang an, Nhat tan, Tay tuu, and Vinch tuy village of Hanoi; also, at Dang hai commune, An hai suburb in Haiphong, Go vap and Hoc mon districts in Ho Chi Minh City and Districts 11 and 12 in Dalat. These areas cover 70-80 per cent of total land area under flowers. Flower production gives high economic return which is 10-20 times more in comparison to rice. Hence, farmers in these areas have better living standards than others.

Major Flower Crops

The major cut flower crops grown in Vietnam and their share.

Table. Major Cut Flowers of Vietnam.

Crop	Area (%)
Rose	40
Chrysanthemum	25
Gladiolus	15
Orchid	10
Others (Carnation, Dahlia, Lilium, Gerbera, Violet, Tuberose, Arum, Peach, Lotus, Camellia, Forget Me Not etc.)	10

Rose is the most popular flower in Vietnam. It can be planted and harvested round the year in the Red River Delta, Dalat, Ho Chi Minh City and other provinces. The popular varieties are the white and red rose in Quang an; light yellow, light pink, dark yellow and dark red rose in Dalat; and black red and creeping rose in other places.

Chrysanthemum is also popular in Vietnam, particularly in the Red River Delta, Dalat, Ho Chi Minh City and other provinces. Popular varieties/species includeChrysanthemum morifolium, C. indicum, C. leacanthemum, Taiwan yellow, Japan yellow, Xuxi chrysanthemum etc. Popular varieties of gladiolus are white, violet, red, light pink, yellow, and short pink.

Most of these flowers are grown in Vietnam in the open fields and only in certain areas are grown in plastic greenhouses, nurseries and experimental gardens. Orchids and Camellia are grown in shadenets for protection from the sun in the summer. But these are not protected against rain, storm and floods, and hence their quality and harvesting are affected.

The country has two different climatic regions *i.e.* tropical in the South and sub-tropical in the North. Besides, in both parts of the country, there are regions with a particular temperate climate, where temperate flowers can be grown such as Sapa and Tamdao in the North and Dalat in the South. Due to this fact, large number of flower varieties could be included in production. These regions are also important for the introduction of new temperate flower varieties, seed production and storage of planting material during summer time, when temperate flowers are difficult to handle.

A tendency of increasing the number of varieties of temperate flowers in production has been observed. These varieties are easily introduced into production because of their commercial value. In recent years a range of varieties of roses, anthurium, chrysanthemum, carnation, orchids etc. were introduced and were quickly established in the local market. As a result, these have become the most important commercial varieties.

Production Systems

As was mentioned earlier, production systems in Vietnam have not yet been developed well; open field is employed for the cut flower industry in most cases. For temperate flowers like rose, chrysanthemum, carnation and lilium, simple shading cover from local material (bamboo) is used; special production system *viz.* greenhouse is yet to be developed. In rare cases in the South small greenhouses are used for cultivation of orchids. But generally orchid production is not popular because of the high investment to build up controlled production systems. Irrigation is provided daily by overhead or surface watering in the summer time. In the rainy season a cover for temperate flowers is needed to avoid development of diseases. At different stages of growth different kinds of organic fertilizers are used.

Due to the fact that producers usually have small area of land, they undertake intensive continuous cultivation. Flower producers therefore practice renewal of the top cultivable layer of soil by removing 10 - 15 cm of top soil and adding new soil every few years. This practice proved to be a good and cheap method to avoid built up of soil-borne diseases, nematodes and mutation exhaustion.

Advances in Production Technology

As mentioned above, cut flower production has become important in Vietnam only in recent years. Research on variety improvement was initiated in the Institute of Agricultural Genetics (AGI) in 1994. It has resulted in releasing a variety of chrysanthemum (CN-93) and collections of local orchids and chrysanthemum. The work on characterization and evaluation of these genetic resources is in progress.

Studies are being conducted to assess the influence of light on blooming of chrysanthemum and the influence of NPK fertilizers on quality of flowers. Official programme on disease control, pest management and post harvest handling has not yet been initiated. However, specific studies are being carried out by some groups of researchers in different insitutions to resolve particular problems.

PRODUCTION OF PLANTING MATERIAL

Planting material is supplied in many cases in the form of exchange between flower growers. Specialized nurseries on production of planting material do not exist. Production of planting material via tissue culture is employed in some cases for roses, carnation, chrysanthemum, authurium and orchids. However, planting material is mainly supplied by the importer or multiplied by conventional methods. There is a great demand for planting material of introduced temperate flower varieties of chrysanthemum, anthurium, carnation and orchids, but an effective system of in vitro production has not yet been developed. The main reason for this situation is the lack of investment and effective technology of in vitro production.

Most of the flower varieties in Vietnam are cultivated from seed (Chrysanthemum, Dianthus caryophyllus, Gladiolus communis), branch (Chrysanthemum, Dianthus caryophyllus) bud, branch and cuttings (Rose, Orchid) and bulbs (Gladiolus communis, Lilium longiflorum Thumb). Propagation from tissue culture is employed for orchids, Dianthus caryophyllus and Chrysanthemum.

Conventional methods are simple, cheap and easy to apply, but virus and other diseases develop easily, which lowers the quality. Seedlings from in vitro methods are strong, but more expensive. Since the market for flowers is limited, farmers are more interested in conventional methods.

MARKETING

Marketing of cut flowers in Vietnam is not organized. There is no organization or association for marketing and distribution of cut flowers; hence it traditionaly follows the sequence of producers Õ businessman Õ market. This situation adversely affects cut flower production.

For certain types of cut flowers the country has a potential to increase production for export but can not realize it. In Vietnam, flower production is largely undertaken for local consumption. Previously, Vietnamese flowers were exported to Russia and East Europe (Chrysanthemum, Gladiolus), though in a small quantity. Presently, there is only one company dealing with export - import.

POTENTIAL FOR CUT FLOWER PRODUCTION DEVELOPMENT

Cut flower production has a good potential for further development in Vietnam, and it appears that the cut flower industry is on the way to being well established in the next few years. The agro-climatic conditions are suitable for cut flower production development. The Government is also encouraging this activity.

CONSTRAINTS IN CUT FLOWER PRODUCTION DEVELOPMENT

The export market has not yet been developed; therefore, there is no motivation to make use of the country's potential. Also, advanced technologies are not available. Moreover, equipment and transport facilities to ensure good quality of flowers are lacking.

CONCLUSION

Vietnam is a country with good potential for cut flower production development. To promote this, it is necessary for Vietnam to collaborate with other countries in the region in order to be able to apply advanced techniques in production, harvesting and storage of cut flowers.

FLOWER FORCING FOR CUT FLOWER PRODUCTION

Flower forcing is an operation or treatment to the plant, after it reaches the ripeness-to-flower stage, in order to stimulate it to flower at a specific date (*e.g.* on New Year's day), or during off-season period.

The flowering date/period may be earlier or later than the normal date/period of flowering.

The goals of flower forcing are off-season production and specific-date production. Cut flowers which are available during the normal season are plentiful, thus fetching a low price. Sometimes the farmers have to sell their produce even at a loss.

In some cases, flowers which could not be sold are either left on the plants or are spoiled after being harvested. Thus, it would be beneficial for farmers to produce cut flowers during the off-season period to obtain higher price, although the inputs may be higher.

Similarly, the demand for cut flowers is generally very high during certain occasions such as New Year, Christmas, Mother's Day, Memorial Day, Valentine Day, Graduation Exercise Day, etc. Thus, it will be to the farmers' advantage if they can produce cut flowers to be available on these specific dates.

The objectives of forcing plants to flower during off-season or at certain specific dates are:

- To avoid surpluses of in-season cut flowers
- To avoid wastage or spoilage of surplus cut flowers
- To avoid danger of epidemics
- To distribute employment throughout the year
- To increase farmers' income
- To reduce imports and trade deficit
- To satisfy customers at the time of their needs

PHYSIOLOGY OF FLOWERING

The process of flowering follows the following sequence:
Proper stimuli (temperature and photoperiod).
Ripeness-to-flower stage · Vernalin · Florigin · Flower.
Factors affecting flowering are photoperiod, temperature and humidity.
Flowering behaviour of plants is controlled by seasonal changes.
There are two types of flowers with respect to the seasonal effect on flowering:

- Little influence of seasonal changes: *e.g.* roses, marigold, chrysanthemum, heliconia, etc.
- Great seasonal influence: *e.g.* jasmine, dendrobium orchids, etc.

Seasonal factors can be of various types viz:

i. *Photoperiodic influence:* This includes short day plants, which are temperature and humidity influenced; also long day plants, which are also temperature and humidity influenced.
ii. *Temperature influence:* This includes low temperature requiring plants, which are photoperiodic and humidity influenced; as also, high temperature requiring plants which are also photoperiodic and humidity influenced.
iii. *Humidity influence:* Including low and high humidity requiring plants.

FORCING OPERATION

Flower forcing can be achieved by adjusting the factors effecting flowering behaviour, *viz.* photoperiod, temperature and humidity. Chemical flower forcing can be done through the application of fertilizers, used both for retarding and stimulating flowering; and plant hormones, including gibberellin, growth retardants and growth inhibitors. Mechanical flower forcing is achieved by operations such as, pruning, leaf trimming, ringing, budding or grafting, smoking, low-temperature storage and breaking dormancy.

FORCING OF SOME CUT FLOWERS IN THAILAND

Dendrobium Orchids

Flowers are available all the year round, but more profuse in rainy season and less profuse in winter (which happens to be the time when the demand is high). Pinching-off flower buds in August/September (to save nutrients for later blooming) and application of special fertilizer (high in P and K) during October/November, 3 to 4 times, induces blooming from November onwards for December/January harvest.

Seasonal flowers are produced during rainy season (June to August), requiring longday condition. No flower develops after September when shortday condition commences. Above ground parts wither and die down, and rhizome enters dormancy period until next rainy season.

Providing additional light breaks dormancy. Most effective is 3 hours of light in the middle of the night. Should be started soon after daylength is shortened (21 September). In this way, plants will continue to produce flowers all the way up to the New Year day, provided enough humidity and nutrients are given.

Marigold

Flowers are available round the year. It is a day neutral plant which takes 60 - 70 days from seeding to harvest. Timing of flowering can be maintained by fixing the date of seeding 60 to 70 days prior of harvest date. The recommended period is 65 days. For example, if the flowers are to be harvested for New Year, seeding should be done on 27 October, transplant on 6 November and pinching on 22 - 24 November. Flower buds emerge by 5 December, and blooming starts from 25 December to 5 January.

Lotus

Flowers are available all the year round, but the plant needs standing water. Low temperature during winter reduces the amount of blooming. Thus, it is recommended that the level of water be reduced to 50 cm in order to raise water temperature. In this way, blooming will be the same as in summer, both in terms of amount and size of flower.

High temperature during the summer speeds up the growth of lotus plant and the blooming time, but flower size remains small. Thus, water level should be raised up to 75 cm (from original 50 cm level in the winter). This way, water temperature will be reduced and the amount of bloom and size of flower will be maintained.

During the rainy season, due to the addition of rain water the level of water may increase, thus it is recommended that the level of water be maintained at 50 cm in order to accommodate additional rain water. The blooming will be maintained as normal.

Chrysanthemum

It is a short day plant, with critical value of 14.4 hours. Thus, it will bloom all the year round under Thai conditions, having maximum daylength of 13.3 hours in June. Daylength can be extended by giving artifical light after sunset for about 3 hours during the early stage of growth to keep seedlings in vegetative stage until one month prior of the planned harvest date. For example, if the planned harvest date is New Year, cutting should be made in September and transplanted to the growing plot for rooting to occur. Seedling is kept under light regime of more than 14.5 hrs by providing artifical light (100 w incandescent bulb) until 1 December (seedlings should be at least 30 cm high). These will bloom on 1 January.

However, as chrysanthemum blooms profusely during the period of low temperature which commences in December, it fetches a low price in the market even during the time of Christmas or New Year. Thus, some farmers avoid producing flowers during such period but shift it to the summer. The problem is that the temperature during summer is quite high and not optimum for chrysanthemum growth.

The most suitable place of production of chrysanthemum flowers is on the highlands of northern Thailand where the temperature during the summer is around 16-20°C. The same principle of flower forcing is applied, but in this case the daylength in northern Thailand may be lower than the critical value for certain varieties. As practiced in Doi Phu Kha of Nan Province, black cloth is used to completely cover the plant house from 16.00 to 08.00 hrs. for 30 days after the cuttings have been exposed to long day conditions (supplemented with artificial light).

Jasmine

Flowers are available thoughout the year, but very profuse during the rainy season and scarce during the winter. To produce jasmine for winter-season harvest, the following operations are recommended:

i. One month before the planned date of harvest, stop watering for 2 - 3 days until the plants show sign of wilting.
ii. Prune the plant to a round shape so that blooming will emerge from mature branches when induced in stage (iii). In this way the blooms will be of large size and healthy as they receive full sunlight.
iii. Apply balanced fertilizer (*e.g.* 15-15-15) at the rate of 30 g/plant and water heavily. Keep watering normally everyday. Flower buds will emerge within 10 days and blooming will occur within 25-30 days after pruning.

Amaryllis

Flowers are available all the year round, although depends on the availability of bulbs which are available from the Netherlands in the summer (June onwards). It takes about one month after the bulbs are planted to have full bloom ready for the harvest of cut flowers. It is recommended to keep the bulbs in a refrigerator (4°C) for at least two months. The bulbs must be planted one month prior to the required date of flowering.

Gladiolus

Flowers can be available all the year round, provided the weather is optimum (requires cool climate). Blooms 90 - 100 days after planting.

In cool climate preheat corms before planting for 2 weeks at 27-32°C. This will force such corms to flower early. In warm climate soak in GA3 solution

(10 - 25 ppm) before planting. This will accelerate flower by hastening differentiation of floral primordia. Growth retardants (*e.g.* CCC) promote initiation of floral primordia (by reducing endogenous GA3 level, or counterbalancing its inhibitory effect on floral initiation.

Roses

Flowers are available round the year but more profuse in the cool season.

For blooming during Christmas - New Year, and Valentine Day, the following procedure is suggested:

- Cut the branches in November. This will stimulate flowering during Christmas - New Year. Flowering will take place 43 days after this pruning. Cutting the flowers on 23 December will further stimulate flower bud initiation to bloom on 10 February. Flowering will take place 49 days after this cutting.
- Pinch (*i.e.* cut only the tip off) instead of pruning as in (i) above, on 10 November for 23 December harvest. This operation automatically stimulates flower initiation for 10 February harvest.

The following operations should be followed:

- Do not pinch healthy flower buds after 10 November as blooming will be due in 30 days. Allow them to bloom for 10 December harvest.
- Do not pinch flower buds which are still very small.
- Pinch all other branches, including those flower buds which are not healthy, deformed, etc. Do not pinch too far down to healthy leaves, unless they are water sprouts which should be cut off from the base.

POTENTIAL OF COMMERCIAL FLORICULTURE IN ASIA

The Asian flower industry has the potential to become the leading flower industry in commercial floriculture worldwide. Looking to the future, the Asian flower industry could rival, if not surpass, the size and scope of the European flower industry which presently dominates global commercial floriculture.

EVOLUTION OF THE WORLD CUT FLOWER MARKET PLACE

Forty years ago, demand for cut flower by consumers around the world was satisfied by local cut flower production. In Europe, per capita consumption was significant, and consumer culture required a large supply of cut flowers for gifts, occasions, and everyday use. As a result, cut flower production in Europe was sizable. Gradually as transportation systems developed throughout this region, it became possible to distribute cut flowers grown in southern areas of Europe to northern areas of Europe. Consequently, the European flower industry began to extend its boundaries for cut flower production and along with this expansion grew the influence of the European flower industry. This background history could be considered the beginning of commercial floriculture as we know it today.

When the world energy crisis occurred in 1973, the marketing plan for distributing cut flowers grown in different European countries to Holland for sale through the Dutch flower auction and back to markets throughout Europe became a significant production opportunity for southern European cut flower growers.

Increasingly larger quantities of cut flowers were grown in southern Europe to meet the demand for cut flower sales through Holland. Flower growers in the southern regions had a price advantage over growers located in northern regions because cut flower production was more expensive for northern growers during the winter season due to increased energy costs required to obtain quality flowers in controlled temperature greenhouses. Then, competition for southern European cut flower growers intensified when Israeli cut flower growers, who were located further south entered the market with product to be sold through the Dutch flower auction.

Israeli growers had the production advantage of being further south where they could produce cut flowers in open fields or plastic tunnels year round, eliminating most of the overhead expenses for greenhouses and heating systems.

But in order to develop a potentially lucrative export cut flower industry for themselves, the Israelis needed to address limiting factors to their success. The two main limiting factors were transportation costs to Europe and a water shortage if production were to expand.

Solutions to these limiting factors were found for Israeli growers. In the case of transportation costs which offset growers cost advantage in terms of energy compared to growers in southern Europe, the government provided transportation subsidies which have reduced the costs to the growers to ship their cut flower product to Europe, thereby maintaining a competitive cost advantage over European growers. As for the water shortage, research on irrigation systems that would conserve water usage was applied to production systems for cut flowers.

Through the 1970's, the activities of the European flower industry had begun to influence cut flower production and sales beyond the borders of Europe. Cut flower sales through the Dutch flower auctions had gained a share of the United States market.

This was achieved by promotion activities in the USA supported by the Holland Flower Council which encouraged Americans to purchase more cut flowers for gifts, occasions and everyday use, similar to consumer habits in Europe. Most of the flowers sold to the USA through the Dutch flower auctions are shipped to the USA by air through New York.

Simultaneously, Miami, USA, was being developed as a key import distribution base for cut flowers being grown in Columbia, South America and shipped north. This caused considerable competition for local cut flower growers in the USA. Manufacturers and suppliers from the European flower industry were quick to find opportunity in this situation. Not only were South American cut flower growers purchasing varieties from Europe but flower growers from the USA were persuaded to invest in production systems and equipment from Europe in hopes of becoming more efficient producers like the Dutch growers who had once faced competition from southern European growers. As a result, the United States flower industry owes a significant share of its growth in terms of promotion and sales and improved production systems to the influence of the European flower industry.

It is worthwhile to mention that the Israeli flower industry has become a formidable competitor of the European flower industry. Israeli cut flower producers ship significant quantities of product into the USA market via both New York and Miami.

This compensates Israeli producers for the reduction in cut flower sales to the European market which is increasingly being supplied by flower growers from regions in Africa, especially Kenya. Also, Israelis have been successful in selling their production equipment and varieties to flower growers in other countries.

Continuing to advance in the 1980's, the European flower industry began seeking further opportunity and expansion in Asia by 1985. Japan's bubble economy was starting to inflate and discretionary income spending by the Japanese was rising. European flower imports made headway into the lucrative market in Japan. Within a few years, as economies in Korea, Taiwan, and Hongkong strengthened, the European flower industry moved into these markets with their cut flower exports as well.

Since the early 1990's, the European flower industry, as a worldwide leader in commercial floriculture, has been impacting the rest of Asia with cut flower imports from Holland and sales of flower varieties, production equipment, and technology for new production operations in Asia. Israeli cut flower producers, manufacturers, and suppliers have followed but, one step behind. The main difference between the European flower industry and the Israeli flower industry is that the European flower industry enters their new markets by launching aggressive marketing campaigns which call attention to the quality and image of Dutch flowers.

These campaigns stimulate demand by new consumers for their cut flower products. So far, the Israelis have not particularly created an image for end consumers of Israeli flowers. This difference is one of the factors which contributes to why the European flower industry is the world wide leader in commercial floriculture.

Initially, commercial floriculture production in southeast Asia was developed because of increasing need for low cost flowers by the European cut flower market place. European flower traders identified commercial floriculture production in southeast Asian countries as a source of supply. Ironically, Dutch auctions often served to re-distribute this product to the Japanese market. By the mid to late 1980's, Dutch importers/exporters had begun selling floriculture product in Japan. With economies expanding, the "little tigers", *i.e.* Taiwan, Korea and Hongkong were the next Asian targets with market needs for floriculture products from Europe and potentially from other Asian countries which could produce floriculture products less expensively.

The development of the commercial cut flower industry in Asia has been unlike that of Israel, African countries, south and central American countries. In the latter regions, cut flowers have been a product produced mainly for export with no thought of a potential domestic market.

On the other hand, in Asia, whereas cut flowers were initially produced for export, the market potential has rapidly changed to include opportunities for supplying the local market as well. This unique development is on account of the rapid strengthening of economies in the region, high population densities, and the consumer perception which has been promoted heavily by the European flower industry that the use of fresh flowers in one's every day life represents an improved, quality lifestyle.

Today and in the future, the potential for commercial floriculture expansion in Asia, including production for domestic and export sales of cut flowers, is greater than ever before. The elements for success needed to transform the Asian flower industry into the worldwide leader of commercial floriculture have been implied by reviewing the evolution of the world cut flower market place.

OPPORTUNITIES FOR CUT FLOWER DEVELOPMENT IN ASIA - MARKETS

There is considerable opportunity in both export and domestic markets for cut flowers from southern Asia. In case of export markets, it seems reasonable to set short, middle, and long term goals with respect to entering these markets. Simultaneously, efforts to develop domestic markets within and among countries in southeast Asia should be undertaken.

GOALS FOR ESTABLISHING EXPORT MARKETS FROM SOUTHEAST ASIA

Short - Term: Countries in Asia - Japan, Korea, Taiwan, Singapore

Already, cut flowers have been exported from southeast Asia to other countries in Asia, especially Japan, Korea, Taiwan, and Singapore. The advantages in marketing product to these countries is that they are relatively near, shipping requirements in terms of postharvest storage have been minimal, though this situation may not persist, and often contacts to find buyers have been made only through relative or previous business relationships.

In the case of Japan, it is well known that custom regulations and quality requirements are among the strictest in the world, though the price fetched for product is high. This incentive is attractive despite the difficulty to achieve success in marketing cut flowers to Japan.

In terms of distribution to these countries, it is important to examine what are the practical transportation routes to these countries from southeast Asia. It could be that developing a transportation hub in Bangkok or Singapore would be convenient for shipping cut flowers from southeast Asian countries, including cut flowers from Yunnan province in China, to countries in northeast Asia. Numerous commercial flights already are in place regarding this possible shipping pattern for cut flowers from the region. The transportation concept of a shipping hub would imply that initially all cut flower production intended for export to northeast Asian destinations would first be shipped to the hub, *e.g.* Bangkok, and then re-distributed to specific destinations in Japan, Korea, Taiwan, and Singapore. This routing for northeast Asia could be extended to Australia and New Zealand.

This centralized shipping hub would require expanding postharvest handling and storage facilities at the airport terminal and increasing freight forwarding activities for cut flowers in Bangkok. At the same time, shipping procedures and storage requirements for cut flowers to be transported from southeast Asian countries would be streamlined by the ability to focus on transportation requirements to Thailand rather than an array of destinations. In the short term, this could serve as a way to shorten the time to improve distribution of cut flower product from all southeast Asian countries to these lucrative northeast Asian markets.

Mid-Term: Countries in North America - Canada and USA; Central Europe

At first consideration, it may seem unlikely that cut flowers exported from southeast Asia could penetrate the markets in North America or central Europe as these markets are already well supplied with cut flowers from other regions such as south and central America and southern Europe, Israel, and Africa. However, in the case of North America, the entry points for cut flower imports are mainly New York and Miami, leaving open the opportunity for establishing an import entry point in Vancouver or San Francisco, especially for cut flowers shipped from a Chinese port, *i.e.* Kunming. Recently, with the return of Hongkong to China, there has been an exodus of Chinese from Hongkong to Canada establishing residence in North America. Already the San Francisco area in the USA is home to many ethnic Chinese.

Due to the high density of Chinese in these two western entry points to North America, it is highly conceivable that Chinese flower importing companies could proliferate if the opportunity were available. This was certainly the case with Miami which already housed a large population from south and central America when flower imports from Columbia, Ecuador, and Costa Rica began to enter the USA market on the east coast. Until now, domestic producers of cut flowers in California on the west coast of the USA have been holding up against imports from south and central America but, would surely retreat if low cost, high quality imports from southeast Asia via China were available. The reason for setting this export goal as "mid-term" is that airline shipping routes from China to these cities are not as well established at this time compared to routes from Bangkok to northeast Asian destinations.

It might be proposed to establish Kunming in China, which is currently being developed as the center for cut flower production in China, as the transportation hub for distributing flowers from southeast Asia to North America. Another mid-term export goal would be to market flowers from southeast Asia to central Europe. This could be achieved, transportation wise also through Bangkok, which already has airline clearances for many central European cities, including Amsterdam.

Another factor in identifying mid-term export market goals is that both north American and central European markets are already well supplied with an abundance of quality, low cost flowers from other regions. These export markets are also both further away in distance so that shipping costs are not particularly an advantage. Marketing factors which will affect the attainment of these mid-term goals will likely be related to price and customer service. The capability for the Asian flower industry to initially offer these export market customers low prices, a full production mix, convenient and reliable shipping service will increase the chances of gaining market share in a market place which already has supply.

Long-Term: Countries in Eastern Europe

Lastly, there will be opportunities for marketing cut flowers from southeast Asia to Eastern Europe, particularly northeastern Europe. This would be another likely destination for flowers being shipped out of Kunming. At the moment, economies in northeastern Europe need to improve before significant quantities of cut flowers would be in demand, but by that time (5-10 years) it is likely that transportation routes will have expanded greatly from China to its neighbouring territory, northeastern Europe. As in the past, no doubt winter production of cut flowers in a southern area will surpass cut flower production in the north.

GOALS FOR ESTABLISHING DOMESTIC MARKETS WITHIN SOUTHEAST ASIA

Similar to establishing export markets for cut flowers from southeast Asia, focusing on improving transportation is a key factor to establishing distribution into domestic markets within and among southeast Asian countries. It is proposed that distribution between these countries may be improved if a transportation hub system were enacted. Also, in the case of developing domestic markets, it is likely that efforts to add or improve wholesale markets in major cities would facilitate greater distribution of cut flowers within these countries.

OPPORTUNITIES FOR CUT FLOWER DEVELOPMENT IN ASIA - PRODUCTION

Opportunities for cut flower development in Asia with respect to production are not location dependent, meaning that any area of southeast Asia should be able to capture significant cut flower production opportunity if pursued. In present and future commercial floriculture markets, production opportunities exist if the goal to reliably produce high quality product in consistent quantities can be attained.

To reliably produce high quality cut flowers in consistent quantities requires optimum production management. Production management strategies for implementation and training of personnel will be different for different types of production operations. Varying types of production operations might be commercial cut flower companies, large or small; government directed agricultural farming communities; or rural development programmes managed by public or private funds.

For each of these types of production operations, dissemination of pertinent information needed to manage production may require different strategies. Therefore, a key factor to realizing production opportunities is education. Ideally, educational resources should be developed and supported by public and private sectors.

An example of educational resources which could be critical to the success of implementing optimum production management for cut flower enterprise in Asia would be a training center/demonstration showcase of production technology adapted to local growing conditions for specific cut flower crops. This educational institution or organization could facilitate dissemination of information to management at cut flower operations or provide training of technical personnel involved in cut flower operations.

Enlisting the support of floriculture professionals from outside and within Asia to coordinate activities in research and education for the purpose of furthering floriculture development in Asia would have significant impact on the pace and scope of cut flower development in Asia. Formally organized, it would be possible to solicit funds through international agencies to support an institution set up for this purpose, *i.e.* International Center for Floriculture in Asia. Departments related to production and marketing management for cut flower crops would be defined in order to reliably produce high quality cut flowers in consistent quantities regardless of the location in southeast Asia.

Such departments could be:

- Cultural production, information and management.
- Postharvest production, information and management.
- Equipment use, design, and engineering.
- Genetic improvement of cut flower varieties.
- Consumer marketing.

If sufficient attention and support were given to attaining the goal of reliably producing high quality cut flower product in consistent quantities among Asian countries, then it would be reasonable to expect that the Asian flower industry would soon overtake any other region's flower industry, including the European flower industry, in production.

Moreover, given the rapidly increasing rate of spending among Asian consumers for cut flowers, it is also reasonable to expect that the Asian flower industry will soon surpass consumption rates for cut flowers compared to that of other regions.

LEADERSHIP IN COMMERCIAL FLORICULTURE WORLDWIDE

It is exciting to discuss how and why the Asian flower industry has the potential to not only "take the lead" in volume and consumption of floriculture products but, to "become the leader" of commercial floriculture worldwide. This may suggest the most important marketing and production opportunities for cut flower development in Asia.

Once again, in review of activities of the European flower industry, it can be said that the European flower industry has led the way in floriculture production, research, floriculture education and training programmes, development of floriculture postharvest handling and storage technology; floral marketing strategies; grades and standards for floral product; and transportation systems for distribution of flowers. Countries within the European region and outside of the region depend on product innovation from Europe to yield new flower varieties. Innovative floriculture production systems have been promoted and exported to other regions of the world. Even floral design concepts have been defined by the European flower industry. The European flower industry exports more than just flowers.

The European flower industry exports flowers, equipment and technology to produce flowers, marketing programmes to increase consumer purchases, and with those marketing programmes, cultural traditions and attitudes about flowers which includes floral design, uses of floral products, *i.e.* decoration, gifts, holidays, and appreciation for flowers.

Celebration of western holidays and the purchases of flowers for these holidays have become a main theme of floral marketing for consumers in Asia. Two examples of this are cut roses for Valentine's Day and cut carnations for Mother's Day. This type comprehensive involvement with floriculture from disseminating the use of production equipment and technology to influencing consumer purchases of flowers by promoting western holidays associated with floral purchases has established the European flower industry as the leader of commercial floriculture worldwide. Let it be asked, could the establishment of an Asian flower industry spread its influence among countries in Asia and the West? Why not? How would this be done?

There are many opportunities for marketing cut flowers from Asia. In addition to developing a new transportation concept for improving global distribution of cut flowers produced in Asia, there are opportunities in developing marketing strategies which would boost global consumption of flowers produced in Asia. First of all, unifying the image of flowers produced in Asia to create an identity of Asian floriculture would be one step. Then, this could be implemented by introducing an Asian calendar of floral holidays which indicated particular flowers and how they are used in association with these holidays.

Even if certain flowers are not presently used in association with a particular holiday or celebration in Asia, there is no reason that floral associations from each country could not invent the concept. For example, in China, on a certain day in the autumn, the moon and all things associated to be round like the moon are appreciated. Millions of round cakes, labeled "moon cakes" are sold to consumers.

Why could not this holiday become a commercial opportunity for the Asian flower industry and pom pom mums become the flower of choice for this day as a gift item along with "moon cakes". If the Japanese celebrate Christmas now with all its commercialism, why could they not celebrate "Moon Day" in the autumn? The same can be said for Americans and Europeans. In Vietnam, flower consumption is auspicious on the 1st and 15th of the month in compliance with Vietnamese traditions. This seems like another convenient way for the Asian flower industry to boost cut flower consumption all over the world twice a month if there were an Asian calendar of floral holidays reminding consumers that it was that time of the month to buy flowers.

Imbedded in this type of floral promotion strategy is the opportunity for the Asian flower industry to develop new varieties of flowers which would characterize these holidays. For example, in case of "Moon Day" not just any type of pom pom mum would qualify, but a particular colour, size, shape, etc. could be developed.

Given the cultural differences among Asian countries, the challenge would be to incorporate as many different cultural traditions representing the different Asian cultures as part of the coordinated Asian flower industry effort. This approach would allow each country to capitalize exclusively on a particular holiday, boosting production and sales for that country according to the holiday celebrated. Once the mission of developing an Asian flower industry was accepted, it would be certain that marketing ingenuity from within Asian flower associations would take over and demand for cut flower products from Asia would soar. Opportunities in production of cut flowers in Asia would go beyond just the need to increase the rate of production. New production systems for efficiency or environmental protection could be developed which would be adapted to conditions of flower production in Asia. Research priorities at flower research institutions throughout Asia could be directed towards developing innovative solutions which supported global sustainability. Then, in a technology turnaround, it would be countries in the west which would be sending exchange staff to Asia to study new flower production systems. Novel engineering or design applications in equipment and supplies which would be used by producers of flowers for the Asian flower industry could be patented and exported to flower producers in other countries outside of Asia. Endless opportunities. Endless opportunities. The establishment of an Asian flower industry could direct the unlimited opportunities for development of cut flower production in Asia and development of worldwide flower markets for cut flowers from Asia.

6

Soil, Water and Nutrient Management

Soil as a medium for plant growth can be described as a complex natural material derived from weathering of rocks and decomposition of organic materials, which provide nutrients, moisture and anchorage for plants. Soil is a mixture of minerals, organic matter (humus), air and water. An ideal soil for plant growth is about 50 percent solids consisting of minerals and organic material. The organic portion consists of residues from plants, animals and other living organisms. Under optimum conditions for plant growth, about half of the space between soil particles — pore space — is filled with water, and the remainder with air. Soil compaction reduces pore space and the amount of air and water the soil can hold, thereby restricting root growth and the ability of plants to take up nutrients from the soil.

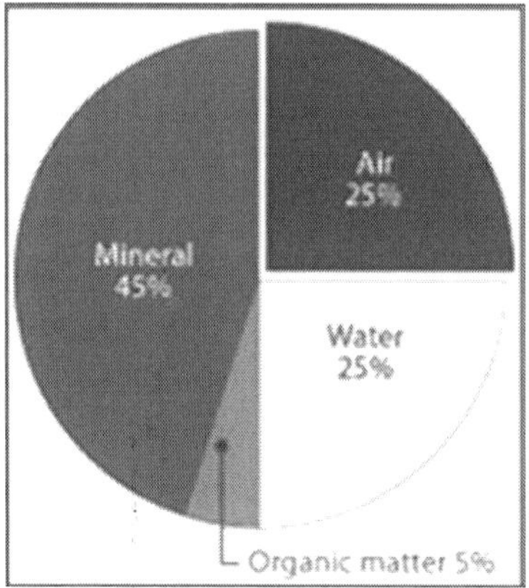

Fig. Volumetric Content of Four Principal Soil Components for an Ideal Soil at Ideal Moisture Content for Plant Growth.

PHYSICAL PROPERTIES

Soil Colour

The colour of soil has little effect on plant growth but is an indicator of soil properties that do affect plant development. Colour is an indicator of organic matter content, drainage and aeration.

- Black
- High in organic matter (4 percent or more).
- Brown
- Good organic matter content and well drained.
- Red
- Low in organic matter, well drained. Red colour is due to the presence of iron (often ferric oxide, Fe_2O_3).
- Gray
- Low in organic matter, poorly drained. Gray colour is due to an excess of water and poor aeration. Gray colour is due to the presence of iron (often ferrous oxide, FeO).
- Yellow
- Low in organic matter, well drained.
- Mottling effects in subsoil
- Iindicates both well and poorly drained conditions during the year due to fluctuations in water table.

Soil Structure

Soil structure refers to the arrangement of soil particles into aggregates. Any physical disturbance influences soil structure. The addition of calcium (Ca), magnesium (Mg) or organic matter improves the structure of soil by enhancing aggregation, the ability of soil particles to hold together as a coherent mixture. Organic matter acts as a bonding agent in holding soil particles together to form aggregates. Excessive sodium (Na) levels in soils cause dispersion of soil particles that can result in poor soil structure. Development of desirable soil structure increases porosity (the amount of pore space in the soil), reduces erodibility and improves water-holding capacity, root penetration and ease of tillage.

Soil Texture

Soil texture refers to the percentage of sand (2.0 to 0.05 mm), silt (0.05 to 0.002 mm) and clay particles (less than 0.002 mm) that make up the mineral portion of the soil. Loam is a variable mix of these three textural classes.

- Sand adds porosity. Silt adds body to the soil. Clay adds chemical and physical properties that affect the ability of the soil to take up nutrients through adsorption to soil particles.
- Soil texture affects the following soil characteristics:
  - Water-holding capacity
  - Nutrient-holding capacity
  - Erodibility
  - Workability
  - Root penetration
  - Porosity

- Soil texture affects soil fertility and nutrient management:
  - Most sulfur deficiencies occur in sandy soils.
  - Nitrogen is easily leached from sandy soils. Loss of soil nitrogen (denitrification) is more common on heavy, clay soils.
  - Potassium can leach from sandy soils but is immobile in medium- to fine-textured soils.

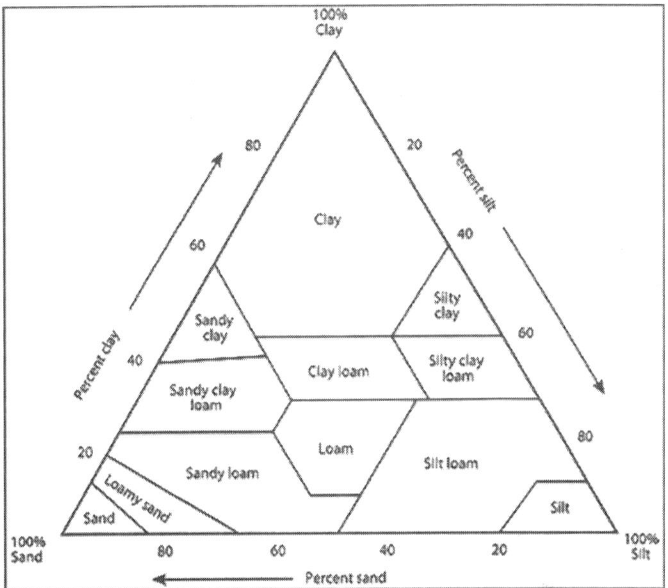

Fig. The Soil Textural Triangle Shows the Percentage of Sand, Silt and Clay in Each of the Textural Classes. Soil Texture can be Measured Accurately in a Laboratory. Soil Texture can also be Estimated in the Field by a Hand-Feel Method.

Table. Soil Texture as Defined by Soil Texture Class and Estimated by a Hand-Feel Method.

Soil textural group	Soil textural class	Feel
Coarse to very coarse (more than 70 percent sand)	Sand, loamy sand	Feels gritty, does not ribbon or leave smear on hand.
Moderately coarse	Sandy loam	Feels gritty, leaves smear on hand, does not ribbon, breaks into small pieces.
Medium	Silt, loam, silt loam	Feels smooth and flourlike, does not ribbon, breaks into pieces about 1/2 inch long or less.
Moderately fine	Sandy clay loam, clay loam, silty clay loam	Forms ribbon that breaks into pieces about 3/4 inch long, sandy clay loam will feel gritty.
Fine (more than 40 percent clay)	Sandy clay, silty clay, clay	Forms long, pliable ribbon more than 2 inches long; sandy clay will feel gritty.

Adsorption vs. absorption :

- Absorption refers to the process of one substance permeating, or passing into the body of, another.
- Adsorption takes place at surfaces and refers to the adhesion of the molecules of one substance at the surface of another.

Soil Organic Matter

Soil organic matter, or humus, is the partially decomposed residue of plants, animals and other organisms. Organic matter refers to all organic material in the soil, including fresh crop residues. Organic matter improves soil structure by acting as a bonding agent that holds soil particles together in aggregates. Without organic matter, aggregates are less stable and can be easily broken apart. Good soil structure promotes water movement and root penetration while reducing soil crusting, clod formation and erosion.

Organic matter provides plant nutrients, mainly nitrogen and sulfur and smaller amounts of phosphorus. About 20 pounds of nitrogen are released by decomposition of every 1 percent of organic matter in the soil. Organic matter is a primary reservoir for available forms of micronutrients (mainly zinc and boron). Soil organic matter also improves the cation exchange capacity of the soil, its ability to hold positively charged molecules, or ions, of mineral nutrients. Soil organisms : Soil organisms vary in size from microscopic bacteria, fungi and algae to those visible to the naked eye, such as earthworms and insects. They perform both beneficial and detrimental functions in the soil.

Microbes decompose organic matter and release nutrients for plant uptake. Bacteria called rhizobia are responsible for fixing atmospheric nitrogen as plant-available forms in root nodules on legumes. Some fungi and nematodes are responsible for plant diseases, and many soil insects damage crops. Fertilizer has positive effects on soil microorganisms by providing more nutrients and increased crop residues. Application of anhydrous ammonia will temporarily reduce populations of microorganisms in the zone of application.

CHEMICAL PROPERTIES

Soil pH

Soil pH is a relative measure of the hydrogen ion concentration (H^+) in the soil. The pH value can vary from a minimum value of 0 to a maximum value of 14.

Optimum soil pH range varies for different kinds of plants:

- Most vegetables and fruits
- pH 6.0 to 7.0
- Potato
- pH 4.8 to 5.5
- Blueberry, rhododendron and azalea
- pH 4.3 to 5.3
- Acidic
- pH less than 7
- Neutral
- pH = 7

- Alkaline
- pH greater than 7

Soil pH affects the availability of nutrients to plants. In acid soils (pH is low) calcium and magnesium become more available to plants, whereas the micronutrients iron, aluminum and manganese become soluble and can reach levels toxic to plants. These micronutrients also can react with phosphorus to form compounds that are insoluble and not available to plants. In alkaline soils (pH is high), several soil micronutrients, including zinc, copper and cobalt, become less available to plants. Also at high pH, phosphorus precipitates (becomes insoluble) with the higher levels of calcium in the soil and therefore becomes less available to plants.

Soil pH affects the population and activity of microorganisms. The activity of nitrogen-fixing bacteria associated with legumes is impaired in acid soils, resulting in less nitrogen fixation.

Several natural processes cause most soils to become more acidic over time:

- Soils formed from acidic rock parent materials like sandstone or shale will be more acidic than those formed from limestone.
- The higher the rainfall, the more leaching of alkaline elements such as calcium and magnesium, leaving acidic elements such as hydrogen, manganese and aluminum. Acid rain can also acidify the soil.
- Nitrogen from fertilizer (ammonium sulfate), decomposing organic matter, and manure as well as nitrogen fixation by legume roots all increase soil acidity.
- Crop removal of nutrients such as calcium and magnesium causes acidity to develop. This effect is more pronounced with legumes than with nonlegumes.

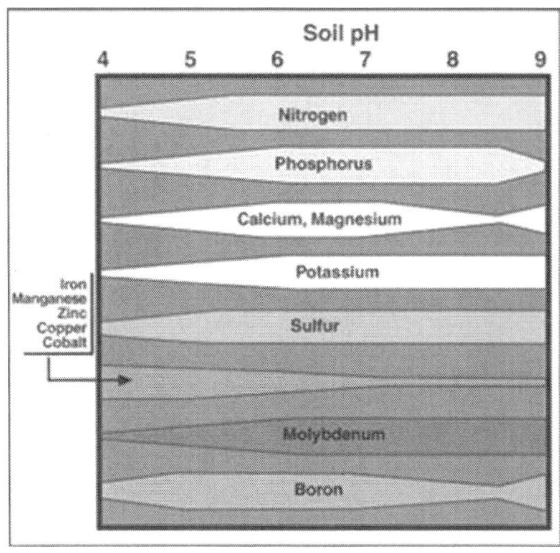

Fig. Soil pH Affects Nutrient Availability to Plants. The Width of the Band Indicates the Relative Availability of Each Plant Nutrient at Various pH Levels.

CORRECTING SOIL ACIDITY

Soil acidity is corrected by adding a liming material to reduce the hydrogen ion concentration and increase the level of alkaline/basic cations (positively charged ions) in the soil. Acid-producing hydrogen ions are adsorbed on exchange sites or present in soil solution. These hydrogen ions are in equilibrium between adsorbed and solution states. The hydrogen ions adsorbed to the cation exchange sites serve as a reservoir for neutralizable/reserve acidity that rapidly replaces the hydrogen ions in the soil solution that are neutralized by lime. A soil pH test tells how acidic a soil is, but it does not measure the neutralizable/reserve acidity.

Thus, to determine how much lime is required to raise the pH, the soil must be tested for neutralizable acidity. This is done in soil testing labs by measuring buffer pH. Two soils with the same pH may have very different amounts of neutralizable acidity and therefore will have different lime requirements. In general, the lime requirement of a soil increases with the content of clay and organic matter.

Liming material	**Composition**	**Relative neutralizing value**
Calcium carbonate	$CaCO_3$	100
Calcitic limestone	$CaCO_3$ + Impurities	50 to 100
Dolomitic limestone	$CaCO_3$ + $MgCO_3$ + Impurities	90 to 109
Quick (burned) lime	CaO	150 to 180
Hydrated (slaked) lime	$Ca(OH)_2$	115 to 135
Ground shells		80 to 90
Wood ashes		40 to 80

- To raise pH, add:
  - Limestone before planting. It will take 3 to 6 months for results.
  - $CaCO_3 + H_2O$ to $Ca + H_2CO_3$
  - $Ca + 2H^+ + CO_3^-$ to $Ca + H_2O + CO_2$
- To lower pH, add:
- Iron sulfate ($FeSO_4$) or aluminum sulfate ($Al_2(SO_4)H_3$) — chemical reaction, quick results.

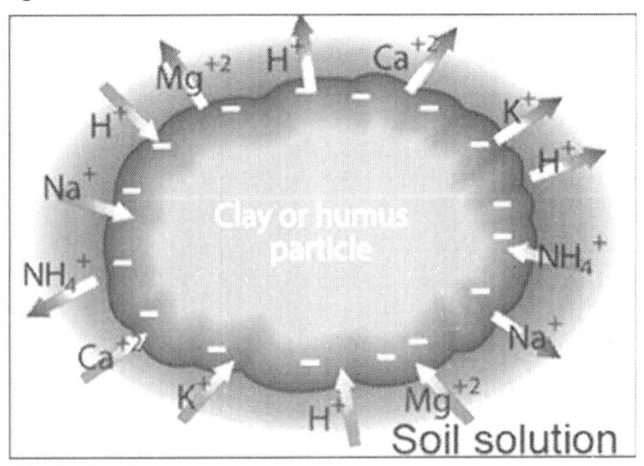

- Elemental sulfur — biological reaction, takes 3 to 6 months to show results.
- Acids (use with caution)
- H_2SO_4 to $2H^+ + SO_4^{-2}$

To determine how much lime your soil needs to correct soil acidity, it is important to test for buffer pH, which indicates neutralizable acidity.

CATION EXCHANGE CAPACITY

Cation exchange capacity (CEC) is a measure of the total amount of exchangeable cations (positively charged ions) a soil can adsorb. Nutrient cations in the soil include positively charged ions such as calcium (Ca^{+2}), magnesium (Mg^{+2}), potassium (K^+), sodium (Na^+) and hydrogen (H^+). In soil tests, CEC is reported in milliequivalents (meq) per 100 grams of soil. The exchangeable cations in the soil are in equilibrium with those in the soil solution (water in the soil). As plants remove nutrients (cations) from the soil solution, they are replenished from the adsorbed cations, which are then available for plant uptake.

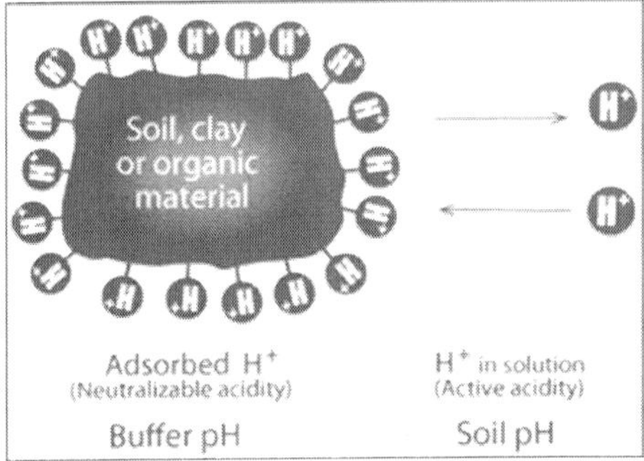

Fig. Exchangeable Nutrient Cations Adsorbed on Soil Particles Exist in Equilibrium with Cations in the Soil Solution. Cations from the Particles Replenish Those Taken Up from the Soil Solution by Plants.

- Soils differ in CEC depending on clay and organic matter content. CEC of clay varies from 4 to 100 meq per 100 g. Humus has an average CEC of 200 meq per 100 g.

Table. The Higher the Clay Content of the Soil, the Greater its Cation Exchange Capacity (CEC).

Soil	CEC (meq/100 g)
Sand	2 to 5
Sandy loam	5 to 12
Loams	10 to 18
Silt and silt clay loams	15 to 30
Clay and clay loams	25 to 40

- Soils with low CEC (1 to 10) have high sand content and low water-holding capacity. They require less lime to correct a given pH, and leaching of nitrogen and potassium is more likely.

- Soils with high CEC (15 to 40) have high clay or humus content and high water-holding capacity. They require more lime to correct a given pH and have a greater capacity to hold nutrients.
- Soil texture affects CEC. The more clay, the higher the CEC.
- The higher the CEC, the more cations a soil can retain.

Anion Retention in Soils

Anions are negatively charged ions. They are retained by positively charged surfaces in the soil, but only in negligble amounts. Negatively charged ions, such as nitrate and phosphate anions, are repelled by clay/humus particles, which are also negatively charged.

For this reason, anions are susceptible to leaching losses in soil solutions:

- Anions such as nitrate (NO_3^-), sulfate (SO_4^-) and chloride (Cl^-) are highly soluble and move with water.
- The phosphate anion (PO_3^-) does not move freely in soils largely because it forms relatively insoluble compounds with iron and aluminum in acid soils (low pH) and with calcium in alkaline soils (high pH).

PLANT NUTRITION

Table. Seventeen Essential Plant Nutrients Derived from Air, Water and Soil.

Plant nutrient	Source		
	Air	Water	Soil
Carbon	X		
Oxygen	X	X	
Hydrogen		X	
Primary nutrients			
Nitrogen			X
Phosphorus			X
Potassium			X
Secondary nutrients			
Calcium			X
Magnesium			X
Sulfur			X
Micronutrients			
Boron			X
Chlorine			X
Copper			X
Iron			X
Manganese			X
Molybdenum			X
Nickel			X
Zinc			X

Seventeen elements are considered essential nutrients for plant growth, and 14 of these elements come from the soil. If there is a deficiency of any essential element, plants cannot complete their vegetative or reproductive cycles. Some of these nutrients combine to form compounds that make up cells and enzymes. Other nutrients are necessary for certain chemical processes to occur.

CONCEPT OF MOST LIMITING NUTRIENT

Just as the capacity of a wooden bucket to hold water is determined by the height of the short stave, crop yields are restricted by the soil nutrient in shortest supply. Increasing the height of the nitrogen (N) stave in the bucket does not increase the bucket's capacity. The unless sulfur fertility is improved, the value of other fertilizer nutrients is reduced. Soil testing discovers the limiting nutrients (short staves) and maximizes fertilizer returns.

Fig. The Most Limiting -Nutrient in a Soil Determines the Growth and Reproduction of Plants.

ESSENTIAL PLANT NUTRIENTS FROM THE SOIL

Nitrogen (N)

Deficiency symptoms :

- Slow growth and stunting.
- Yellow-green colour leaves.
- "Firing" of tips and margins of leaves; yellowing begins with mature leaves.

Fertilizer Sources

The first number on bag of inorganic fertilizer (e.g., 20-10-20) refers to the amount of pure nitrogen per 100 pounds of fertilizer:

- Calcium nitrate (15.5-0-0)
- Potassium nitrate (13-0-14)
- Sodium nitrate (16-0-0)
- Ammonium nitrate (34-0-0)
- Ammonium sulfate (21-0-0)

- Diammonium phosphate (21-53-0)
- Urea (45-0-0)

NITROGEN

Nitrogen is a building block of plant proteins. It is an integral part of chlorophyll and is a component of amino acids, nucleic acids and coenzymes. Most nitrogen in the soil in tied up in organic matter. It is taken up by plants as nitrate (NO_3^-) and ammonium (NH_4^+) ions from inorganic nitrate and ammonium compounds. These compounds can enter the soil as a result of bacterial action (nitrogen fixation), application of inorganic nitrogen fertilizer, or conversion of organic matter into ammonium and nitrate compounds.

Not all nitrates in the soil are taken up by plants. Nitrates can be leached beyond the root zone in sandy soils or converted to nitrogen gas in wet, flooded soils. Nitrogen fixation by soil microbes immobilizes nitrogen, making in available for later use by plants. A soil test is the best way to determine how much nitrogen fertilizer should be added to your soil. Application rates for specific crops are based on typical yield goals, the organic matter content of the soil, the previous crop produced on that soil, and the amount of manure used.

PHOSPHORUS (P)

Deficiency symptoms :

- Slow growth and stunting.
- Purplish colouration on foliage of some plants.
- Dark green colouration with tips of leaves dying.
- Delayed maturity.
- Poor fruit or seed development.

Fertilizer sources

The second number on a bag of inorganic fertilizer (e.g., 20-10-20) refers to the percentage of P_2O_5 by weight:

- Superphosphate (0-20-0)
- Concentrated/Treble superphosphate (0-46-0)
- Monoammonium phosphate (12-62-0)
- Diammonium phosphate (21-53-0)

PHOSPORUS

Plants use phosporus to form the nucleic acids DNA and RNA and to store and transfer energy. Phosphorus promotes early plant growth and root formation through its role in the division and organization of cells. Phosphorus is essential to flowering and fruiting and to the transfer of hereditary traits. Phosphorus is adsorbed by plants as $H_2PO_4^-$, HPO_4^{-2} or PO^{-3}, depending upon soil pH. The mobility of phosphorus in soil is low, and deficiencies are common in cool, wet soils. Phosphorus should be applied to fields and gardens before planting and should be incorporated into the soil. This is especially important for perennial crops. Application rates should be based on soil testing.

POTASSIUM (K)

Deficiency symptoms :

- Tip and marginal "burn" starting on mature leaves. Lower leaves turn yellow.
- Weak stalks and plants lodge easily.
- Small fruits or shriveled seeds.
- Slow growth.

Fertilizer Sources

The third number on a bag of inorganic fertilizer (e.g., 20-10-20) refers to the percentage of K_2O by weight:

- Potassium nitrate (13-0-14)
- Potassium chloride (0-0-60)
- Potassium sulfate (0-0-50)

POTASSIUM

Potassium is necessary to plants for translocation of sugars and for starch formation. It is important for efficient use of water through its role in opening and closing small apertures (stomata) on the surface of leaves. Phosphorus increases plant resistance to diseases and assists in enzyme activation and photosynthesis. It also increases the size and quality of fruits and improves winter hardiness. Plants take up potassium in the form of potassium ions (K^+). It is relatively immobile in soils but can leach in sandy soils. Potassium fertilizer should be incorporated into the soil at planting or before. Application rates should be based on a soil test.

CALCIUM (CA)

Deficiency symptoms :

- "Tip burn" of young leaves — celery, lettuce, cabbage.
- Growing point dieback. Death of growing points (terminal buds). Root tips are also affected.
- Stunted root growth.
- Premature shedding of blossoms and buds.
- Weakened stems.
- Water-soaked, discoloured areas on fruits — blossom-end rot of tomatoes, peppers and melons; bitter pit or cork spot of apples and pears.

Fertilizer sources :

- Lime ($CaCO_3$)
- Hydrated lime ($Ca(OH)_2$)
- Dolomitic lime ($CaMg(CO_3)_2$)
- Gypsum ($CaSO_4$)
- Superphosphate ($CaHPO_4$)
- Calcium nitrate ($CaNO_3$)

CALCIUM

Calcium provides a building block (calcium pectate) for cell walls and membranes and must be present for the formation of new cells. It is a constituent of important plant carbohydrates, such as starch and cellulose. Calcium promotes plant vigour and rigidity and is important to proper root and stem growth.

Plants adsorb calcium in the form of the calcium ion (Ca^+). Calcium needs can be only determined by soil test. In most cases calcium requirements are met by liming the soil. Potatoes are an exception; use gypsum (calcium sulfate) on potatoes to avoid scab disease if calcium is needed. Gypsum provides calcium to the soil but does not raise the pH level of the soil. Keeping pH low helps prevent growth of the bacteria that cause scab disease.

MAGNESIUM (MG)

Deficiency symptoms :

- Interveinal chlorosis (yellowing) of older leaves.
- Curling of leaves upward along margins.
- Marginal yellowing with green "Christmas tree" area along midrib.
- Generally supplied when soils are limed with dolomitic lime.
- Magnesium deficiencies are most likely to occur on acid, sandy soils.
- Soil tests can be used to determine Mg needs.
- Mg deficiency can be induced by high K applications.

Fertilizer sources :

- Dolomitic limestone ($CaMg(CO_3)_2$) (found in hard water)
- Magnesium sulfate ($MgSO_4$) 10 percent Mg

MAGNESIUM

Magnesium is a component of the chlorophyll molecule and is therefore essential for photosynthesis. Magnesium serves as an activator for many plant enzymes required for sugar metabolism and movement and for growth processes. Plants take up magnesium as the Mg^{+2} ion.

SULFUR (S)

Deficiency symptoms :

- Young leaves are light green to yellowish in colour. In some plants, older tissues are also affected.

- Small and spindly plants.
- Retarded growth and maturity.

Fertilizer sources :

- Elemental ground sulfur (S)
- Gypsum ($CaSO_4$)
- Potassium sulfate (K_2SO_4)
- Ammonium sulfate $(NH_4)_2SO_4$
- Iron sulfate ($FeSO_4$)
- Sulfuric acid (H_2SO_4)

SULFUR

Sulfur is a constituent of three amino acids (cystine, methionine and cysteine) that play an essential role in protein synthesis. Sulfur is present in oil compounds responsible for characteristic odours of plants such as garlic and onion. It is also essential for nodule formation on legumes.

Plants take up sulfur in the form of sulfate (SO_4^{-2}) ions. Sulfur can also be adsorbed from the air through leaves in areas where the atmosphere has been enriched with sulfur compounds from industrial wastes. Sulfur is susceptible to leaching, and sulfur deficiencies can occur in sandy soils low in organic matter. Sulfur needs can be only determined by a soil test.

ZINC (ZN)

Deficiency symptoms :

- Decreasing in stem length and a rosetting of terminal leaves.
- Reduced fruit bud formation.
- Dieback of twigs after the first year.
- Mottled leaves and interveinal chlorosis.

Fertilizer sources :

- Zinc sulfate ($ZnSO_4{\cdot}H_2O$)
- Zinc oxide (ZnO)
- Zinc chloride ($ZnCl_2$)
- Chelating agents EDTA, HEEDTA and NTA help make certain nutrients more available to plants.

ZINC

Zinc is an essential component of several enzymes in plants. It controls the synthesis of indoleacetic acid, an important plant growth regulator, and it is involved in the production of chlorophyll and protein. Zinc is taken up by plants as the zinc ion (Zn^{+2}).

Zinc deficiencies are more likely to occur in sandy soils that are low in organic matter. High soil pH, as in high-lime soils, the solubility of zinc decreases and it becomes less available. Zinc and phosphorus have antagonistic effects in the soil. Therefore zinc also becomes available in soils that are high in phosphorus. Wet and cold soil conditions can cause zinc deficiency because of slow root growth and slow release of zinc from organic matter.

IRON (FE)

Deficiency symptoms :

- Interveinal chlorosis of young leaves. Veins remain green except in severe cases.
- Twig dieback.
- In severe cases, death of limbs or plants.

Fertilizer sources :

- Ferrous sulfate ($FeSO_4 \cdot H_2O$)
- Ferrous ammonium sulfate[$Fe(NH_2)_2(SO_4)_2 \cdot 6H_2O$]
- Chelating agents EDTA, HEEDTA and NTA help make certain nutrients more available to plants.

IRON

Iron is taken up by plants as ferrous ion (Fe^{+2}). Iron is required for the formation of chlorophyll in plant cells. It serves as an activator for biochemical processes such as respiration, photosynthesis and symbiotic nitrogen fixation. Turf, ornamentals and certain trees are especially susceptible to iron deficiency, although in general, lack of iron in the soil is not a problem. Symptoms of iron deficiency can occur on soils with pH greater than 7.0. Specific needs for iron can be determined by soil test, tissue test and visual symptoms.

MANGANESE (MN)

Deficiency symptoms :

- Interveinal chlorosis of young leaves.
- Gradation of pale green colour with darker colour next to veins. No sharp distinction between veins and interveinal areas as with iron deficiency.

Fertilizer sources :

- Manganous sulfate ($MnSO_4$)
- Manganese oxide (Mn_2O_3)
- Manganous oxide (MnO)

MANGANESE

Manganese serves as an activator for enzymes in plant growth processes, and it assists iron in chlorophyll formation. Plants obtain this nutrient from the soil in the form of manganous ion (Mn^{+2}). Manganese deficiency in soils is not common but can occur in sandy soils with a pH of 8. Soil pH is a good indicator of manganese availability, which can increase to toxic levels in highly acidic soils (pH less than 4.5). Crops most responsive to manganese are onions, beans, potato, spinach, tomato, peas, raspberries, strawberries, apples and grapes.

COPPER (CU)

Deficiency symptoms :
- Stunted growth.
- Dieback of terminal shoots in trees.
- Poor pigmentation.
- Wilting and eventual death of leaf tips.

Fertilizer sources :
- Copper sulfate ($CuSO_4 \cdot 5H_2O$)
- Cupric oxide (CuO)
- Cuprous oxide (Cu_2O)
- Chelating agents EDTA, HEEDTA and NTA help make certain nutrients more available to plants.

COPPER

Copper is an activator of several enzymes in plants. It may play a role in production of vitamin A. Deficiency interferes with protein synthesis. Copper deficiencies are not common in soils. Plants take up copper from the soil in the form of cuprous (Cu^+) or cupric (Cu^{+2}) ions. Crops most responsive to copper are carrots, lettuce, onions and spinach.

BORON (B)

Deficiency symptoms :
- Death of terminal buds, causing lateral buds to develop and producing a "witches broom" effect.
- Thickened, curled, wilted and chlorotic leaves.
- Soft or necrotic spots in fruit or tubers.
- Reduced flowering or improper pollination.

Fertilizer sources :
- Granular borax ($Na_2B_4O_7 \cdot 10H_2O$)
- Solubor ($Na_2B_8O_{13} \cdot 4H_2O$)

BORON

Boron regulates the metabolism of carbohydrates in plants. It is essential for the process by which meristem cells (cells that divide) differentiate to form specific tissues. With boron deficiency, plant cells may continue to divide, but structural components are not differentiated.

Boron is taken up by plants as the borate ion (BO_3^-). Plants differ in their boron needs. Plants with high boron requirements are cauliflower, broccoli, turnip, brussels sprouts, apples, celery and alfalfa. Boron can be limiting on sandy soils low in organic matter. Do not overapply, because boron toxicity can occur (*e.g.*, beans). Soil testing for boron can predict fertilizer requirement.

MOLYBDENUM (MO)

Deficiency symptoms :

- Stunting and lack of vigour. Similar to nitrogen deficiency due to the key role of molybdenum in nitrogen use by plants.
- Marginal scorching and cupping or rolling of leaves.
- "Whiptail" of cauliflower.

Fertilizer sources :

- Sodium molybdate ($Na_2Mo_4 \cdot H_2O$)
- Ammonium molybdate [$(NH_4)_2Mo_2O_{24} \cdot 4H_2O$]

MOLYBDENUM

Molybdenum is taken up by plants as molybdate ions (MoO_4^-). Molybdenum is an essential micronutrient that enables plants to make use of nitrogen. Without molybdenum, plants cannot transform nitrate nitrogen to amino acids and legumes cannot fix atmospheric nitrogen.

Molybdenum deficiency can occur in acidic, sandy soils. Liming the soil to pH 6 will correct the problem. Soil applications, foliar applications or coating seed with molybdenum are also effective. Cauliflower is the main vegetable crop sensitive to low levels of molybdenum in the soil.

CHLORINE (CL)

Deficiency symptoms :

- Wilting followed by chlorosis (yellowing).
- Excessive branching of lateral roots.
- Bronzing of leaves.
- Chlorosis and necrosis (tissue death) in tomatoes and barley.

Fertilizer sources :

- Calcium chloride ($CaCl_2$)
- Ammonium chloride (NH_4Cl)
- Potassium chloride (KCl)

CHLORINE

Chlorine is required in photosynthetic reactions. Deficiency of chlorine in soils is rare because of its universal presence in nature. Plants take up chlorine as chloride ion (Cl^-).

NICKEL (NI)

Deficiency symptoms

Deficiency occurs only in plants growing in solution culture, such as hydroponics, not in soil. Nickel deficiency in plants causes accumulation of urea in leaf tips because of depressed urease activity in leaves. This accumulation of urea causes necrosis of leaf tips.

Fertilizer sources

Because nickel deficiency does not occur in soils, there are no reported sources of nickel fertilizer for soil.

NICKEL

Nickel is taken up by plants as Ni^{+2}. Nickel is a component of the enzyme urease, which is needed to prevent toxic accumulations of urea, a product of nitrogen metabolism in plants. Nickel is thought to participate in nitrogen metabolism of legumes during the reproductive phase of growth. It is also essential for seed development. High levels of nickel in the soil can induce zinc or iron deficiency by competition between these elements in plant uptake.

SOIL TESTING FOR HEALTHIER LAWNS AND GARDENS

The soil test is an excellent gauge of soil fertility. It is an inexpensive way to maintain good plant health and maximum productivity without polluting the environment by overapplication of nutrients.

Soil fertility fluctuates throughout the growing season each year. The quantity and availability of mineral nutrients are altered by the addition of fertilizers, manure, compost, mulch, lime or sulfur and by leaching. Furthermore, large quantities of mineral nutrients are removed from soils as a result of plant growth and development and the harvesting of crops. A soil test will determine the current fertility status. It also provides the information needed to maintain optimum fertility year after year.

Some plants grow well over a wide range of soil pH, while others grow best within a narrow range of pH. Most turf grasses, flowers, ornamental shrubs, vegetables and fruits grow best in slightly acid soils (pH 6.1 to 6.9). Plants such as rhododendron, azalea, pieris, mountain laurel and blueberries require a more acidic soil to grow well.

A soil test is the only precise way to determine whether the soil is acidic, neutral or alkaline. A soil test takes the guesswork out of fertilization and is extremely cost effective. It not only eliminates the expense of unnecessary fertilizers but also eliminates overuse of fertilizers and helps to protect the environment.

When is the best time for a soil test? Soil samples can be taken in the spring or fall for established sites. For new sites, soil samples can be taken anytime when the soil is workable. Most people conduct their soil tests in the spring. However, fall is a preferred time to take soil tests if one suspects a soil pH problem and wants to avoid the spring rush. Fall soil testing will allow you ample time to apply lime to raise the soil pH. Sulfur should be applied in the spring if the soil pH needs to be lowered.

How to take a soil sample? Most errors in soil testing occur when the sample is taken. Potential sources of errors include the following:

- Too few cores per sample
- Failure to properly divide the area to be sampled
- Failure to cover the whole area
- Contaminated sample

Taking a representative sample is important in soil testing. Use a trowel, spade and sampling tube/core samplers:

- For garden and lawn establishment or renovation, take a 6-inch sample.
- For established lawns, take a 3- to 4-inch sample after removing thatch.
- Sample from five or more scattered/random spots in the test area.

What soil sampling tools do I need? A soil sample is best taken with a soil probe or an auger. Samples should be collected in a clean plastic pail or box. These tools help ensure an equal amount of soil to a definite depth at the sampling site. However, a spade, knife, or trowel can also be used to take thin slices or sections of soil. Push the tip of a spade deep into the soil and then cut a 1/2-inch to 1-inch slice of soil from the back of the hole. Be sure the slice goes 6 inches deep and is fairly even in width and thickness. Place this sample in the pail. Repeat five or six times at different spots over your garden. Thoroughly mix the soil slices in the pail. After mixing thoroughly, take out about 1-1/2 cup of soil and mail or, preferably, take it to your University Extension center. You can also mail or deliver it to the MU Soil and Plant Testing Laboratory in Columbia or at the Delta Research Center in Portageville. It is important that you fill out the soil sample information form completely and submit it with your sample. By indicating on the form the crops you wish to grow, you can get specific recommendations. How often should I test my soil? Soil should be tested every two to three years. In sandy soils, where rainfall and irrigation rates are high, samples should be taken annually. What tests should be run? In general a regular fertility test is sufficient. This includes measurement of pH, neutralizable acidity (NA), phosphorus, potassium, calcium, magnesium, organic matter (OM) and cation exchange capacity (CEC).

What do the test result numbers mean?

Some labs report soil test values as amounts of available plant nutrients, and others report extractable nutrients that will become available to the plants. Fertilizer rates are given in pounds of actual nutrient (as distinct from pounds of fertilizer) to be applied per 1,000 square feet.

APPLY FERTILIZERS AS RECOMMENDED BY SOIL TEST

All fertilizer recommendations given in a soil test report are based on the amount of nutrient (N, P_2O_5, K_2O) to apply for a given area. Lawn and garden recommendations are given in pounds per 1000 sq. ft. From the given recommendations it is necessary to select an appropriate fertilizer grade and determine how much of this fertilizer to apply to the garden area. Numbers on fertilizer bags indicate the exact percentages of nutrients by weight: 100 lb of 5-10-10 fertilizer contains 5 lb of nitrogen (N), 10 lb of phosphate (P_2O_5), and 10 lb of potash (K_2O). Because it is difficult to achieve the exact amount of all recommended nutrients from the garden fertilizer blends available in the market, it is important to match the nitrogen requirement.

Example: A soil test recommendation for your vegetable garden calls for 2 lb of N/1000 sq. ft, 0 lb of P_2O_5 /1000 sq. ft and 1 lb of K_2O. The garden is 40 ft by 10 ft.

- Step 1
- Calculate the area to be fertilized. Multiplying length by width, the area of the garden is 40 x 10 = 400 sq. ft.
- Step 2:
- Select the fertilizer to be used. Match the ratio of nutrients recommended to the fertilizer grades available. The N-P-K nutrient ratio based on the soil test is 2-0-1. Ideally, a fertilizer such as 10-0-5 or 20-0-10 or 30-0-15 should be selected. At the local garden store, fertilizer bags marked 20-10-10, 27-3-3 and 25-0-12 are available. The one marked 25-0-12 best matches the ratio of 2-0-1 recommended by soil test.
- Step 3
- Determine the fertilizer amount to apply: Divide the recommended amount of nutrient by the percentage of the nutrient (on a decimal basis) in the fertilizer.
  - First calculate the fertilizer recommendation for the garden area:
  - 2 lb of N/1000 sq. ft x 400 sq. ft/ garden = 0.8 lb of N per 400 sq. ft garden. 100 lb of the 25-0-12 garden fertilizer blend will have 25 lb of N and 12 lb of K_2O.
  - To provide 0.8 lb of N for the 400 sq. ft garden you would require:
  - 100 lb for fertilizer blend /25 lb of N x 0.8 lb of N = 3.2 lb of the fertilizer blend required to provide the N requirement of the garden. Since the fertilizer blend ratio is almost the same as the recommended ratio, it will provide the required amount of K (1.6 lb of K_2O) to the garden.

Note

The weight of 2 cups of dry fertilizer is approximately 1 pound. Therefore to meet the garden fertilizer recommendation, you will need about 6 cups of the fertilizer blend (25-0-12) material for the 400 sq. ft area.

Recommended application rate for various granular fertilizers to apply one pound of nitrogen.

	Application rate		
	Per 1000 square feet		Per 10 square feet
Source	Pounds	Cups	Tablespoons
10-10-10	10	20	4
8-8-8	12.5	25	5
12-4-8	8	16	3
16-4-8	6	12	2
20-10-10	5	10	2
12-6-6	8	16	3

NUTRIENT MANAGEMENT

Nutrient management is managing crop fertility inputs and other production practices for efficient crop growth and water quality protection. Nutrient management plans for site-specific situations minimize undesired environmental effects while optimizing farm profits and production.

Nutrient management planning (NMP) is a Best Management Practice or BMP. The term "nutrient management" is most often associated with animal manure management, but applies to all fertility crop inputs whether organic materials, livestock production by-products, or inorganic commercial fertilizers.

All BMP's are research-proven, commonsense, doable, and economical guidelines for applying manures and fertilizers in crop production. They should be followed anytime field applications are being made. What is Nutrient Management Planning? Nutrient management planning principles are basic, sound fundamentals necessary for good business management.

Nutrient management is:

- Knowing what you have
- Knowing what you need
- Managing wisely
- Documenting your management

Nutrient management plans must be site-specific, tailored to the soils, landscapes, and management objectives of the farm. In effect, NMP is much like developing a cash-flow analysis using pounds of N and P instead of dollars.

Steps of Nutrient Management Planning :

1. Obtain accurate soil information for each field or management unit. This could require a new farm soil map or adaption of existing NRCS maps. Soil samples should be obtained and analyzed according to recognized soil fertility analytical procedures.
2. Estimate yield potential for each field based on soil productivity and intended management. It is impossible to foretell growing seasons, but average yields over last 5 to 7 years should provide a reasonable estimate. It is very, very important to be realistic.
3. Calculate plant nutrients required to reach the yield potential. Nutrient uptake and removal data for common crops are available from the

NRCS, the local Extension office, and several links below. It is important to distinguish between uptake, or use by the growing crop, and removal, or the physical displacement of the nutrients from the field in the harvest.

4. Determine the plant-available nutrients in any livestock by-product amendments. The best method is to sample the manures to be used on the field. Table values are available, but accurate nutrient content of manure is site/animal/diet/management specific. Instructions on sampling and analysis of animal manures are available from the local Extension office.
5. Estimate residual nutrient contributions from fertilizer or manures applied in previous seasons. Manures are the original slow-release nitrogen fertilizer. Usually about 50 per cent of the nitrogen content is available to growing plants the first year following application. Subsequent use is usually a sliding scale. The NRCS NMP Field Office Technical Guide has one such scale for use.
6. If necessary, use a environmental risk assessment tool, the Mississippi Phosphorus Index (PI), to determine the potential for offsite movement of nutrients on a field-by-field basis. The PI incorporates site specific soil conditions and applied BMP's in the evaluation process. Soil test phosphorus levels, soil permeability, field slopes, litter application rates, distance to surface water, and other factors are used to determine the probability of nutrient movement in the landscape. If the PI rating is low, NM may be based on crop nitrogen needs. If the PI is medium, additional BMP's may need to be utilized. If the PI shows a high potential risk for P movement in the landscape, NM should be based on crop P requirements.
7. Apply animal manures and/or commercial fertilizers to supply nutrients when needed by the growing crops using Best Management Practices.
8. Keep records of nutrient sources, application dates, rates, and methods.

ASSISTANCE IN NUTRIENT MANAGEMENT PLANNING

Nutrient management plans are required as part of the general environmental permitting process for various animal operations in Mississippi. Additionally, operations without livestock, but use animal manures whose transport is subsidized by a cost-share programme may be required to prepare a NMP. For these required plans, NMP's developed according to the Practice Standards of the Mississippi Natural Resources Conservation Service are acceptable. These formal plans may be developed by local agents of the Natural Resource Conservation Service or other state approved alternative providers of conservation planning services such as crop consultants or professional engineers. Nutrient management plans may be developed for other uses by the planners listed above or other trained personnel. Individual farmers may develop NMP's to utilize in their own management system or to meet lender requirements.

Nutrient Management Planning Summary:

- Know the soils and fields of your farm
- Be real about what you can produce
- Determine the nutrients you will remove
- Find out what you have available from this year's application
- Calculate nutrients available from previous applications
- Assess environmental risk of nutrient movement
- Use common sense in putting it out
- Keep relevant field records

Best Management Practices for Nutrient Management Using Manure:

- Use soil testing to assess fertility status and follow the recommendations.
- Determine nutrient and moisture content of the manure.
- Use commonsense, attainable yield goals.
- Use the Phosphorus Index to determine application guidelines.
- Rotate fields receiving manure to avoid nutrient buildup and maximize nutrient utilization.
- Use additional fertilizers only when manure nutrients do not meet crop yield goals.
- Incorporate manure when possible.
- Calibrate application equipment.
- Avoid applying manure on wet soils to minimize compaction, runoff and leaching/denitrification.
- Avoid surface application of manure near surface waters.
- Use grass filter strips along ditches and waterways to reduce soil erosion, runoff and nutrient losses.
- Manage overall water flow to maximize water time on the soil surface before drainage.
- For annual crops, apply manure as close to the time of crop utilization as possible.
- Utilize fall cover crops to minimize soil erosion and runoff and to maximize nutrient utilization from manure applications.

Nutrient Management Plan

A crop nutrient management plan is a tool that farmers can use to increase the efficiency of all the nutrient sources a crop uses while reducing production and environmental risk, ultimately increasing profit. It is generally agreed that there are ten fundamental components of a Crop Nutrient Management Plan. Each component is critical to helping analyze each field and improve nutrient efficiency for the crops grown.

Fig. Manure Spreader.

These components include:

- *Field Map:* The map, including general reference points (such as streams, residences, wellheads etc.), number of acres, and soil types is the base for the rest of the plan.
- *Soil Test:* How much of each nutrient (N-P-K and other critical elements such as pH and organic matter) is in the soil profile? The soil test is a key component needed for developing the nutrient rate recommendation.
- *Crop Sequence:* Did the crop that grew in the field last year (and in many cases two or more years ago) fix nitrogen for use in the following years? Has long-term no-till increased organic matter? Did the end-of-season stalk test show a nutrient deficiency? These factors also need to be factored into your plan.
- *Estimated Yield:* Factors that affect yield are numerous and complex. A field's soils, drainage, insect, weed and disease pressure, rotation and many other factors differentiate one field from another. This is why using historic yields is important in developing yield estimates for next year. Accurate yield estimates can dramatically improve nutrient use efficiency.
- *Sources and Forms:* The sources and forms of available nutrients can vary from farm-to-farm and even field-to-field. For instance, manure fertility analysis, storage practices and other factors will need to be included in a nutrient management plan. Manure nutrient tests/ analysis are one way to determine the fertility of it. Nitrogen fixed from a previous year's legume crop and residual affects of manure also effects rate recommendations. Many other nutrient sources should also be factored into this plan.

- *Sensitive Areas:* What's out of the ordinary about a field's plan? Is it irrigated? Next to a stream or lake? Especially sandy in one area? Steep slope or low area? Manure applied in one area for generations due to proximity of dairy barn? Extremely productive—or unproductive—in a portion of the field? Are there buffers that protect streams, drainage ditches, wellheads, and other water collection points? How far away are the neighbors? What's the general wind direction? This is the place to note these and other special conditions that need to be considered.
- *Recommended rates:* Here's the place where science, technology, and art meet. Given everything you've noted, what is the optimum rate of N, P, K, lime and any other nutrients? While science tells us that a crop has changing nutrient requirements during the growing season, a combination of technology and farmer's management skills assure optimum nutrient availability at all stages of growth. No-till corn generally requires starter fertilizer to give the seedling a healthy start.
- *Recommended timing:* When does the soil temperature drop below 50 degrees? Will a N stabilizer be used? What's the tillage practice? Strip-till corn and no-till often require different timing approaches than seed planted into a field that's been tilled once with a field cultivator. Will a starter fertilizer be used to give the seedling a healthy start? How many acres can be covered with available labour (custom or hired) and equipment? Does manure application in a farm depend on a custom applicator's schedule? What agreements have been worked out with neighbors for manure use on their fields? Is a neighbor hosting a special event? All these factors and more will likely figure into the recommended timing.
- *Recommended methods:* Surface or injected? While injection is clearly preferred, there may be situations where injection is not feasible (*i.e.* pasture, grassland). Slope, rainfall patterns, soil type, crop rotation and many other factors determine which method is best for optimizing nutrient efficiency (availability and loss) in farms. The combination that's right in one field may differ in another field even with the same crop.
- *Annual review and update:* Even the best managers are forced to deviate from their plans. What rate was actually applied? Where? Using which method? Did an unusually mild winter or wet spring reduce soil nitrate? Did a dry summer, disease, or some other unusual factor increase nutrient carryover? These and other factors should be noted. It's easier to make notes throughout the year than to remember back six to 10 months.

When such a plan is designed for animal feeding operations (AFO), it may be termed a "manure management plan." In the United States, some regulatory agencies recommend or require that farms implement these plans in order to prevent water pollution. The U.S. Natural Resources Conservation Service (NRCS) has published guidance documents on preparing a comprehensive nutrient management plan (CNMP) for AFOs. The International Plant Nutrition Institute has published a 4R plant nutrition manual for improving the management of plant nutrition. The manual outlines the scientific principles behind each of the four R's or "rights" and discusses the adoption of 4R practices on the farm, approaches to nutrient management planning, and measurement of sustainability performance.

NUTRIENT RECYCLING FOR SUSTAINABLE AGRICULTURE

The total territory of Viet Nam is 32 924 000 ha but only 35 percent of it is used for agriculture. Total cultivated soil is 11 569 591 ha, of which more than 50 percent are problem soils such as arenosol, thionic fluvisol and acrisols.

SOIL FERTILITY STATUS

Beside two alluvial soils of Viet Nam (Red river fluvial soil and Mekong river fluvial soil), soil fertility in Viet Nam is not very high. The widespread soil in Viet Nam has low pH, low C, low N and very low CEC. It is especially true for soil with light texture as sandy soil or acrisol. The dominant feature of a degraded soil (Acrisol). Results of routine soil testing conducted recently reveal that almost all of Vietnamese soil are low in N content, 80 percent of soil

Table. Main Cultivated Soils of Viet Nam.

No.	Soil group name	FAO-Unesco	Area (ha)	Percent		
1	Sandy soil	Arenosol	533 434	4.61		
2	Saline soil	Salic fluvisol	971 356	8.39		
3	Acid sulfate soil	Thionic Fluvisols	1 863 128	0.49	52		
2	pH KCl		4.05	0.23	16		
4	Density	g/cm^3	2.60	9.08	16		
6	Texture			9.96	16		
	0.2-0.02 mm	%	32.44	%	14.05	10.83	16
7	OC	%	1.19	3.88	53		
9	Ca^{++}	cmol$_c$/kg	1.75	0.36	50		
11	K^+	cmol$_c$/kg	0.15	0.11	0.63	44	
14	H^+	cmol$_c$/kg	0.06	0.06	53		
16	0.05	53					
17	K_2O	%	0.19	29.13	52		

samples are deficient in K, 87 percent in P, 72 percent in Ca and 48 percent in Mg (Nguyen van Bo et al., 2003).

FOOD PRODUCTION IN VIET NAM

Rice dominates Vietnamese agriculture. Viet Nam's agriculture sub-sector, particularly rice, has seen a dramatic production increase since 1986. Rice production was 11.6 millions tonnes in 1980, more than 31 million tonnes in 2000 and 38 millions tonnes in 2004. Viet Nam is now in safe situation in terms of food and each year exports about 3 millions tonnes of rice.

Table. Area and Production Figures for the Major Crops in Viet Nam (Statistical Data of Viet Nam Agriculture, Forestry and Fishery 2003, Statistical Publishing House).

Crop	Sown area (thousand ha)	Average yield (tonne/ha)
Rice	7 499.3	4.63
Maize	909.8	3.22
Sweet potato	219.9	7.24
Vegetables	200.0	20-100
Sugarcane	306.4	53.91
Cassava	371.9	14.7
Soybean	166.5	1.35
Groundnut	242.8	1.67
Fruit crops	719.8	6.38
Coffee	513.7	1.50
Tea	116.2	0.22
Rubber	436.5	0.18

Increased production, particularly in rice, has been achieved through greater cropping intensity including fertilizer use rather than any other factor in cropping area.

FERTILIZER USE

Contributing to the increased yield were the widespread adoption and use of improved, high yielding varieties, better irrigation, plant protection, together with inputs of fertilizer. According to research from NISF (1998-2000), one unit of N-P-K gives in average of 7.5-8.5 kg of paddy. Each year, Viet Nam spends almost $600 millions US for importing almost 100 percent, 80 percent and 10 percent of the country'— requirement in K, N, and P respectively, for food production.

Details of fertilizer sources and uses are shown in Tables 4 and 5. Viet Nam farmers used an average of 234 fertilizer unit per ha in 2004 comparison with 134 unit in 1998 and only 18 unit in 1976.

WARNING OF OVERLOADING NUTRIENT TO SOILS

Nutrients sources can be of mineral or organic in origin. The principle of balance fertilization is to make full use of both of these sources in an integrated way that is economically and environmentally sound. Recently, organic farming has been promoted as an alternative to common agriculture in Viet Nam. The adverse effect is that some fertilizers and organic materials contain not only nutrients but also heavy metals. This may be toxic in some level with high accumulation through time. In some areas of very intensive cropping systems with cash crop such as vegetable and flower, fertilizer use were especially high, both for inorganic and organic materials, soil was probably overloading of fertilization.

Table. Mineral Fertilizers and Farmyard Manure using in Different Cropping in North Viet Nam (2003-2004).

Cropping system	Location, Code	NPK (kg/ha)	FYM (tonne/ha)
Rice-Rice-Soybean	HaTay, PT01	22	
Rice-Rice	ThaiBinh, KX04	19	
Rice-Rice-Maize	NamDinh,VB02	17	
Rice-Rice-Maize	NinhBinh, NB02	28	
Maize-Soybean-Maize	VinhPhuc, VT02	8	
Mulberry	VinhPhuc, YL03	8	
Vegetables	VinhPhuc, ML1	110	
Flower	VinhPhuc, ML1	160	

Table. Annual Nutrient Budget (Mean, std) at the Field Levels.

Cropping systems	Input levels	N (kg/ha)	P(kg/ha)	K (kg/ha)	Cu (g/ha)	Zn (kg/ha)
Dry vegetable	Fertilizer + chicken manure + rice straws + compost	390 (350)	291 (203)	3.96 (1.23)		
Water vegetable	Waste water from city + horn + borne + ash + compost	489 (375)	137 (366)	2.43 (3.88)		

Table. Heavy Metals in Some Manure Recycling for Agriculture in North Viet Nam.

	Cu (mg/kg dry)	Pb (mg/kg dry)	Zn (mg/kg dry)	Cd (mg/kg dry)
Sludge	45	27	154	0.39
Std	*5.1*	*0.35*	*5.43*	*0.01*
Pig manure	250	14	556	0.65
Std	*413*	*6.8*	*835*	*0.22*
Goat manure	19	7.5	75	0.21
Std	*2.0*	*2.6*	*12*	*0.05*
Chicken manure	64	16	222	1.49
Std	*54*	*8*	*106*	*1.78*
Cow manure	67	20	153	0.48
Std	*71*	*8*		

INTEGRATED NUTRIENT MANAGEMENT

Soil is a fundamental requirement for crop production as it provides plants with anchorage, water and nutrients. A certain supply of mineral and organic nutrient sources is present in soils, but these often have to be supplemented with external applications, or fertilizers, for better plant growth. Fertilizers enhance soil fertility and are applied to promote plant growth, improve crop yields and support agricultural intensification.

Fertilizers are typically classified as organic or mineral. Organic fertilizers are derived from substances of plant or animal origin, such as manure, compost, seaweed and cereal straw. Organic fertilizers generally contain lower levels of plant nutrients as they are combined with organic matter that improves the soils physical and biological characteristics. The most widely-used mineral fertilizers are based on nitrogen, potassium and phosphate.

Optimal and balanced use of nutrient inputs from mineral fertilizers will be of fundamental importance to meet growing global demand for food (International Food Policy Research Institute, 1995). Mineral fertilizer use has increased almost fivefold since 1960 and has significantly supported global population growth — Smil (2002) estimates that nitrogen-based fertilizer has contributed an estimated 40 per cent to the increases in per-capita food production in the past 50 years. Nevertheless, environmental concerns and economic constraints mean that crop nutrient requirements should not be met solely through mineral fertilizers. Efficient use of all nutrient sources, including organic sources, recyclable wastes, mineral fertilizers and biofertilizers should therefore be promoted through Integrated Nutrient Management (Roy et al, 2006).

DESCRIPTION:

The aim of Integrated Nutrient Management (INM) is to integrate the use of natural and man-made soil nutrients to increase crop productivity and preserve soil productivity for future generations (FAO, 1995a). Rather than focusing nutrition management practices on one crop, INM aims at optimal use of nutrient sources on a cropping-system or crop-rotation basis. This encourages farmers to focus on long-term planning and make greater consideration for environmental impacts.

INM relies on a number of factors, including appropriate nutrient application and conservation and the transfer of knowledge about INM practices to farmers and researchers. Boosting plant nutrients can be achieved by a range of practices covered in this guide such as terracing, alley cropping, conservation tillage, intercropping, and crop rotation. Given that these technologies are covered elsewhere in this guidebook, this section will focus on INM as it relates to appropriate fertilizer use. In addition to the standard selection and application of fertilizers, INM practices include new techniques such as deep placement of fertilizers and the use of inhibitors or urea coatings (use of area coating agent helps to retart the activity and growth of the bacteria responsible for denitrification) that have been developed to improve nutrient uptake.

Key components of the INM approach include:

1. Testing procedures to determine nutrient availability and deficiencies in plants and soils. These are:
    a. Plant symptom analysis – visual clues can provide indications of specific nutrient deficiencies. For example, nitrogen deficient plants appear stunted and pale compared to healthy plants
    b. Tissue analysis and soil testing – where symptoms are not visible, post-harvest tissue and soil samples can be analysed in a laboratory and compared with a reference sample from a healthy plant
2. Systematic appraisal of constraints and opportunities in the current soil fertility management practices and how these relate to the nutrient diagnosis, for example insufficient or excessive use of fertilizers.
3. Assessment of productivity and sustainability of farming systems. Different climates, soil types, crops, farming practices, and technologies dictate the correct balance of nutrients necessary. Once these factors are understood, appropriate INM technologies can be selected
4. Participatory farmer-led INM technology experimentation and development. The need for locally appropriate technologies means that farmer involvement in the testing and analysis of any INM technology is essential (Box 1).

On-farm Testing of Integrated Nutrient Management Strategies in Eastern Uganda

"An action research project carried out by CIAT (Centro Internacional de Agricultura Tropical) in three villages in Eastern Uganda implemented participatory on-farm testing of farmer-designed INM strategies during a two-year process. Twenty farmers representing three soil fertility management classes in the three villages were chosen by the farmer groups as test farmers for intensive monitoring of the on-farm experimentation.

During the diagnostic phase of the PLAR process farmers analysed soil fertility management diversity and resource endowment resulting in the identification and prioritisation of 12 soil fertility and management constraints. Drought was the main constraint, followed by lack of knowledge and skills on soil fertility management, low inherent soil fertility, and soil-borne diseases and pests. The high cost of inorganic fertilizers was ranked number sixth, while soil erosion and poor tillage methods were ranked seventh. During the planning phase, farmers were taken on a farmer exchange visit to meet other farmer innovators who practise some of the proposed technologies.

Farmers designed 11 experiments and they proposed data collection procedures for monitoring and evaluation. Soil samples were collected for laboratory analysis and plant growth was monitored for germination percentage, crop performance, weed management, pests and disease incidence, time of harvesting, and crop yield

Results

Application of farmyard manure at 10 t/ha fresh weight tended to improve maize grain yield in the two years of the project. Although the grain yield increases were not significant, farmers were ready to adapt the technology at large-scale. However, the availability, quantity and quality of the manure in the area is a major constraint to wide-scale adoption of this technology. The farmers designed an experiment to evaluate various sources of phosphorous fertilizers.

There were five treatments or different mixes, including a control with no fertilizers. There was significant response to the various sources of phosphate fertilizers on maize grain yield. However, capital constraints were identified as limiting factors affecting further adoption of this technology. Green manure application did not significantly improve maize yields however the mean annual dry matter (biomass) yields were significantly different. Farmers in the test area have been using green manure for more than five years. Therefore it was proposed that this technology be disseminated without any further on-farm testing.

Farmer evaluation of on-farm experiments shows that simple, inexpensive technologies requiring little labour and locally available resources have a high potential for adoption. However, bio-economic modelling studies showed that a substantial improvement in the socio-economic environment is needed to give farmers sufficient incentives to adopt more sustainable land management practices. The results support the hypothesis that systematic learning with stakeholders, and farmers perceiving economic incentives, are necessary for changing farming practices. However, the capacity of different farmers to invest in improving soil fertility management depends on access to labour, livestock, land, capital and cash at the household level. The options available to poor farmers are much more constrained than those available to the well endowed farmers who are able to invest in large-scale use of organic and inorganic sources of nutrients."

Harsh climatic conditions are a major cause of soil erosion and the depletion of nutrient stocks.

By increasing soil fertility and improving plant health, INM can have positive effects on crops in the following ways:

- A good supply of phosphorous, nitrogen and potassium has been shown to exert a considerable influence on the susceptibility or resistance of plants towards many types of pests and diseases
- A crop receiving balanced nutrition is able to explore a larger volume of soil in order to access water and nutrients. In addition, improved root development enables the plant to access water from deeper soil layers. With a well-developed root system, crops are less susceptible to drought
- Under increasingly saline conditions, plants can be supplemented with potassium to maintain normal growth
- With appropriate potassium fertilisation, the freezing point of the cell sap is lowered, thus improving tolerance to colder conditions

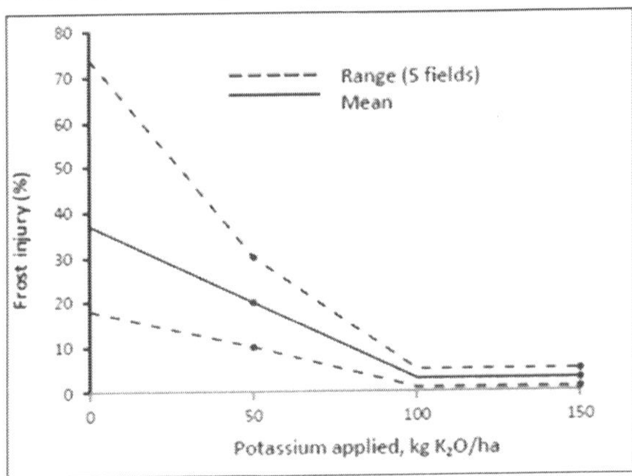

Fig. Effect of Potassium Application on Frost Injury to Potato Crop (Source: Grewal and Sharma, 1978).

ADVANTAGES OF THE TECHNOLOGY TOP:

INM enables the adaptation of plant nutrition and soil fertility management in farming systems to site characteristics, taking advantage of the combined and harmonious use of organic and inorganic nutrient resources to serve the concurrent needs of food production and economic, environmental and social viability.

INM empowers farmers by increasing their technical expertise and decision-making capacity. It also promotes changes in land use, crop rotations, and interactions between forestry, livestock and cropping systems as part of agricultural intensification and diversification.

DISADVANTAGES OF THE TECHNOLOGY TOP:

As well as facilitating adaptation to climate change in the agriculture sector, the INM approach is also sensitive to changes in climatic conditions and could produce negative effects if soil and crop nutrients are not monitored systematically and changes to fertilizer practices made accordingly. In Africa, high transport costs in land-locked countries contribute to prohibitively high fertilizer prices (FAO, 2008b). In the case of small-scale farmers these costs may represent too high a proportion of the total variable cost of production thus ruling out inorganic fertilizer as a feasible option.

FINANCIAL REQUIREMENTS AND COSTS TOP:

The main cost associated with Integrated Nutrient Management relates to the purchase and distribution of inorganic fertilizers which are affected by a range of factors

Table. Average Cost of Fertilizers Per Metric Ton in Africa.

Country	Factors affecting cost	Cost
Mozambique	Coastal country Private market dominated by a single importer making low-volume purchases Absence of retail network resulting in low provision to rural areas Very high transportation costs and poor road infrastructure No local manufacturing or blending facilities Low fertiliser demand and consumption	$554
Malawi	Land locked country Fertilisers make up one of the four largest markets in the country. Net importer with some local production. Government plays central role in importation and delivery through public tender Choice of port (South Africa, Tanzania or Mozambique) greatly affects cost and availability High transport costs due to high fuel prices and poor road infrastructure Subsidised fertiliser programme with farmer voucher scheme Excessive importer, wholesale, and retail margins	$495
Ghana	Coastal country All fertilisers imported Privatised market dominated by three major importer- wholesalers Well-organised distributors and dealers No direct import duties or sales tax Market in growth phase Market price competition robust Predominantly ship (from international suppliers) and truck transport (from Nigerian suppliers and to distributors) High storage costs at ports High inland transport costs	$ 386

Organic fertilizers provide a low-to-no-cost technology for improving soil fertility as long as they can be produced and used within a relatively close distance.

INSTITUTIONAL AND ORGANISATIONAL REQUIREMENTS TOP:

The success of INM will depend upon the combined efforts of farmers, researchers, extension agents, governments, and NGOs. Simply providing fertilizers is not enough to support INM implementation. Appropriate policy frameworks are essential, as are market structures, infrastructure development, credit facilities and the transfer of technology and knowledge. INM requires knowledge of what is required by plants for optimum level of production per cent in what different forms and at what different timings and how these requirements can be integrated to obtain highest productivity levels within acceptable economic and environmental limits. Determining this information will require localised research but will also benefit from the cooperation of national and international agricultural research centres. Extension staff who are able to translate research data into practical recommendations will need to take account of both farmers' expertise and applicable research results. Available knowledge will need to be summarised and evaluated economically in order to provide practical guidelines for the adoption of INM by farmers that have a range of investment capacities.

BARRIERS TO IMPLEMENTATION TOP:

An insufficient availability of credit at an affordable price is frequently mentioned as a constraint on fertilizer use. Access to mineral fertilizer may be limited in rural or underdeveloped areas due to high import prices and high transport costs. A lack of adequate infrastructure for distribution and conservation can also present a barrier for access and use. In addition, fertilizers have a limited shelf-life and may be in high demand (leading to shortages) in peak seasons if appropriate planning is not put in place. Competition for organic resources may be high in areas where crop residues are used for fuel and animal feed.

OPPORTUNITIES FOR IMPLEMENTATION TOP:

A largely untapped source of potential fertilizer is urban waste. Although the quality or fertilizer produced from urban waste does not compare to commercially produced fertilizer, the sludge (Residual, semi-solid material left from industrial wastewater, or sewage treatment processes) contains nitrogen, phosphorous, potassium and other micro-nutrients.

Utilising urban waste for agricultural lands near urban centres puts to good use a material that otherwise would be disposed via costly means (Gruhn et al, 2000). Farmers associations and extension services provide an opportunity for production and dissemination of information on the most cost-effective and appropriate technologies.

SOIL FERTILITY FOR FIELD-GROWN CUT FLOWERS

Field production of cut flowers, particularly annuals, requires many of the same soil fertility management practices as production of vegetables and other outdoor crops.

The main goal is to maintain proper fertility to produce a good yield of high quality flowers for summer sales. Another goal is to maintain soil health and avoid environmental contamination by using fertilizers responsibly.

MAJOR NUTRIENTS

Nitrogen. Adequate nitrogen (N) is critical for cut flower production. A N deficiency will result in poor plant growth, a reduction in flower yield, and the appearance of foliar chlorosis. Too much N may result in too much vegetative growth, cause weak stems, and delay flowering. Nitrogen is the most difficult nutrient to manage because it is required in the greatest amount by plants and is easily lost from the soil by leaching or as a gas. For these reasons N fertilizer must be applied each year to maintain adequate levels for production. Leaching of N as nitrate (NO_3) can be a serious problem not only as a way of losing N for the crop, but also because it can pollute groundwater and surface bodies of water. Too much N fertilizer, high rainfall or excess irrigation, and sandy soil are all factors favouring NO_3 leaching.

Nitrogen fertilization also effects soil pH. Over time the continuous use of ammonium or urea fertilizers lowers the pH of the soil. The acidity is created because ammonium or urea are quickly converted to NO_3 by soil bacteria and acidity is a byproduct of this process. Neutralizing the acidity generated by N fertilizers is one of the reasons for using limestone.

Phosphorus. The effects of phosphorus (P) on cut flower production are less obvious than the effects of N. A P deficiency may result in smaller plants and shorter flowering stems, but foliar symptoms may not be apparent. Phosphorus is often said to be needed for flowering, but the exact effects of P on cut flower crops has not been studied.

Less P than N is needed by plants and P is more stable than N in the soil. Some agricultural soils in Massachusetts test "very high" or "excess" in P because of the regular use of NPK fertilizers over many years. Generally, the potential harm caused by high P is not to the plants, but to the environment. Too much P can lead to runoff to surface bodies of water. This encourages the growth of too much algae and other aquatic plants which leads to a serious decline in water quality. Ideally nutrients should be applied based on need as determined by a soil test. Unlike N, an application of P may not be needed every year. Superphosphate is a water-soluble P fertilizer and is best used to correct a low P condition before planting. Rock phosphate is not water-soluble and it is best used for the long-term maintenance of P in the soil.

Potassium. Potassium (K) deficiency causes marginal chlorosis and burning on the lower leaves first. Unless the deficiency is very serious it may not effect the leaves on the flowering stems, but it could reduce overall yield. Potassium deficiency is most likely to occur on sandy soils low in clay and organic matter. Soils lacking in clay and/or organic matter have a low cation exchange capacity (CEC) and have little ability to retain the cations (*i.e.*, positively-charged ions) such as K, as well as calcium (Ca) and magnesium (Mg).

It may not be necessary to apply K fertilizer to some clay soils and soils high in organic matter every year. To the author's knowledge there is no harm to the environment from excess K and no direct harm of high K to most plants. However, too much K can depress the uptake of Ca and Mg, sometimes to the point that deficiency symptoms of these elements develop.

SOIL TESTING

A yearly soil test for pH and nutrient levels should be made in either the fall or spring Since P, K, Ca and Mg are fairly stable in the soil a fall test gives a good reading on the status of these elements for next spring. If needed, limestone or rock phosphate can he applied in the fall giving these materials a head start in reacting with the soil before spring. Since the status of N is very changeable, a spring test or one right before N application is best.

Lime Requirement

Checking pH and adjusting it by liming is important in managing soil fertility. The general acceptable pH range for flower crops is 5.5 to 7.0, but some plants may have particular optimum pH levels for best performance. pH affects the availability of trace elements and the activity of many beneficial soil microorganisms. Phosphorus availability is highest between pH 6 and 7, so liming an acid soil may help free some plant-available P.

In addition to raising the pH of acid soils limestone serves as fertilizer for Ca and Mg. As a soil becomes acid it also supplies less and less Ca and Mg. Regular use of N fertilizers hastens the formation of acid soil as discussed earlier. Use finely ground dolomitic limestone or calcitic limestone with 5-10 per cent Mg to get both elements.

Ideally pH should be checked once a year either in the fall or the spring. If the pH is not in the desired range a lime requirement test (buffer pH) can be rim to determine how much lime is needed. These tests take into account the soil texture- sandy soils need much less limestone to cause a pH change than soils high in clay or organic matter. The table below can help determine the approximate amount of limestone to use if the soil texture is known. Approximate amount of limestone needed to change soil pH

Table. Pounds of Limestone Per Acre (1 acre = 43,560 sq. ft.).

Change in pH desired 6-8"deep	Sand	Sandy loam	Loam	Silt loam	Clay loam
4.0 to 6.5	2,600	5,000	7,000	8,400	10,000
4.5 to 6.5	2,200	4,200	5,800	7,000	8,400
5.0 to 6.5	1,800	3,400	4,600	5,600	6,600
5.5 to 6.5	1,200	2,600	3,400	4,000	4,600
6.0 to 6.5	600	1,400	1,800	2,200	2,400

FERTILIZING CUT FLOWERS

The goal of fertilizing outdoor cut flowers is to provide adequate levels of nutrients for vegetative growth early in development, strong stems, and a good yield of flowers.

Annuals. Annual species are started from either seed or transplants. Germinating seeds and young seedlings are very sensitive to fertilizer injury so the amount of fertilizer put on initially should be very small. Starter fertilizer should be applied at a rate of about ¼ - ½ lb. of actual N/1000 sq. ft. at seeding either by broadcasting or banding. Bands should be placed about 2" below and to the sides of the seeds. More fertilizer would be applied later after the young plants are well established.

If annual transplants are used to start the crop about 20-30 per cent of the total N requirement can be applied shortly before planting assuming the transplants are healthy and have been hardened-off. Another one-third of the total should be applied as a sidedress when the transplants are about 8-10" tall. The final application should be made with about 4-6 weeks of harvest left to go. Can one application of N fertilizer at transplanting grow a good crop? "Yes" is probably the answer in many cases. However, the split application method is a more efficient means of applying N fertilizer in fields receiving heavy irrigation or rainfall or that have sandy, well-drained soil. Split application provides a more consistent supply of nutrients than one large application at planting.

Perennials. In general fertilizer is applied to encourage shoot development when growth begins in the spring. Approximate amount of limestone need to change soil pH and to support the flowers as they develop. After flowering or harvest most perennials continue to grow building the root system and expanding the crown or other overwintering structures. An additional application of fertilizer is needed to support preparations for winter. So the total fertilizer requirement might be divided as follows: a small application in the spring as growth begins, a sidedress application as the plants approach flowering, and another application after harvesting ends.

Rates of application. In his book on field grown cut flowers, Stevens (1998) recommends fertilizer be applied at 1 to 2 lbs. actual N per 1000 sq. ft. for a diverse group of annuals and perennials. At the 2 lb. rate this would be 20 lb. of 10-10-10 fertilizer. Some large, and fast-growing annuals like sunflower may need more and it can be added as a sidedress if necessary. As in greenhouse practice complete NPK fertilizers are applied to outdoor cuts according to the N requirement. Ideally, however, P and K should only be applied if their need is indicated by a soil test. So rather than using 10-10-10 a grower could use single element carriers such as ammonium nitrate, superphosphate, and potassium chloride to fulfill the exact needs according to a soil test.

COVER CROPS

Once a cut flower crop has been harvested there may be some growing time left in the late summer and early fall. Rather than leaving the soil fallow this would be a good time to grow a cover crop. Cover crops absorb residual nutrients left in the soil, add organic matter, may add N (legume), and may help protect the soil over winter. Below is information from the UMass Extension Vegetable Programme about cover crops:

Winter rye. Winter rye has been an old stand by. It can germinate and make quite a bit of growth, even if planted as late as October. Winter rye is efficient at taking up left over nitrogen. It remains green over the winter and resumes growth early in the spring. It adds little organic matter if plowed under in early spring while still small. If allowed to grow until late may, it can reach three to four feet and contribute a fair amount of organic matter. Unless plowed under while quite small, it can be difficult to break up the clumps of winter rye, making it difficult to seed crops.

Winter rye is a popular cover crop, but others can be used starting in the summer. Oat. This is a very useful cover crop for us in Massachusetts. Unlike winter rye, oat will grow very vigourously and straight up when seeded in the summer. We frequently seed this cover crop during the summer months at the Research Farm in South Deerfield on land that is not in production. Seeds are readily available, but beware that you do not get oats that have been cooked for use as animal feed. Oat seeded in the summer will probably produce viable seed before frost. For this reason you might want to incorporate it (or simply mow it) before it goes to seed. Seeding rate- 100 lbs/acre.

Sorghum x sudangrass. As the name implies this cover crop is a cross between sudangrass and forage sorghum. One nice aspect of this cover crop is that the seed produced is sterile so there is no concern of it going to seed. When seeded in mid-June it can get eight feet tall! There has been some concern about how to manage it in the fall. Experience with this cover crop has been that it is not difficult to manage, however you may want to seed a little of it the first time in order to get some experience with it. It will die with a hard frost.

You can either leave it and till the following spring or incorporate it after a frost and seed rye. Seeding rate- 50 lbs/acre.

Hairy vetch. Hairy vetch is a legume cover crop that has been very successful when planted in late July or early August. This would be a good crop to grow on land that will be in production early next spring. When vetch is seeded in the summer it will not usually survive the winter (whereas it will survive the winter when seeded in late August or September). In the spring the N from the dead vetch will become available, so this is when we want the cash crop in the ground to take advantage of it. It can be seeded alone or in combination with oat. Seeding rates: 25-30 lbs/acre (using a grain drill); 3 5-40 lbs/acre (broadcast); 40 lbs/acre of oat when seeded with vetch. (Author's note: Hairy vetch should be seeded no later than September 15 in Western Massachusetts. If hairy vetch is started in August it may add as much as 100 lb. N/acre when plowed down in the spring.)

7

Common Diseases of Flower Crops

DISEASE OF ROSE

BLACK SPOT (DIPLOCARBON ROSAE)

Symptoms:

- Black lesions with feathery margins surrounded by yellow tissue are found on the leaves. Infected leaves drop prematurely.
- Purple/red bumpy areas on first year canes may be evident.
- Plants may be weakened due to defoliation and reduced flower production may be observed.

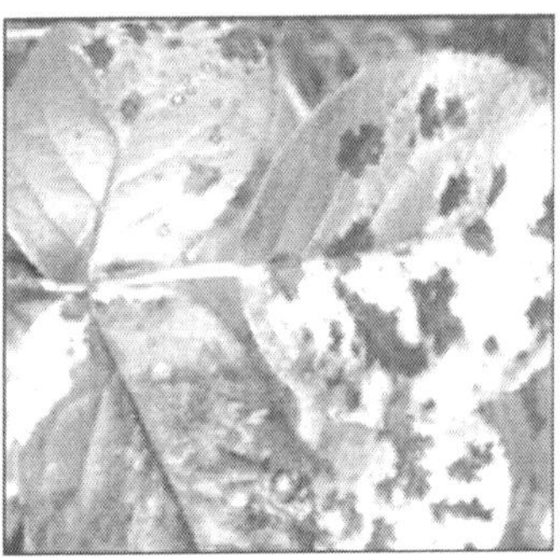

Management:

- Use of Resistant Varieties.
- Cultural-Roses should be planted where the sun can quickly dry the night's dew.
- Space roses far enough apart for good air circulation Avoid overhead watering and keep foliage as dry as possible.
- Sanitation-Remove infected canes and burn diseased leaves.

POWDERY MILDEW (SPHAEROTHECA PANNOSA)

Symptoms:

- The symptom appears as grayish-white powdery substance on the surfaces of young leaves, shoots and buds.

- Infected leaves may be distorted, and some leaf drop may occur.
- Flower buds may fail to open, and those that do may produce poor-quality flowers.
- It can occur almost anytime during the growing season when temperatures are mild (70 to 80 °F), and the relative humidity is high at night and low during the day.
- It is most severe in shady areas and during cooler periods

Management:

- Collection and burning of fallen leaves.

DIE BACK (DIPLODIA ROSARUM)

Symptoms:

- Drying of twigs from tip down wards.
- Blackening of the twigs.
- The disease spreads to root and causes complete killing of the plants.

Management

1. Pruning should be done so that lesions on the young shoots will be eliminated. Apply chaubatia pastic in the pruned area.

RUST (PHRAGMIDUM MUCRONATUM)

Symptoms:

- Damage to lemon yellow pustules appear on lower surface of the leaves and stems. Then the colour changes to blackish red.
- The affected leaves turn yellow deformed and fall prematurely.
- Die back symptom also appear due to weakening of the plant.

Management

Collection and burning of fallen leaves.

DISEASE OF JASMINE

PHYLLODY

Symptoms

Leaves become small malformed and bushy. In the place of flowers green leaf like malformed flowers are formed. Management Selection of cuttings from healthy plants.

DISEASE OF CHRYSANTHEMUM

RUST (PUCCINIA CRYSANTHEMI)

Rust is a serious disease especially in the early spring:

- The disease symptoms are in the form of brown blister-like swellings, which appear on the undersides of leaves.
- These burst open releasing masses of brown, powdery spores. Severely infected plants become very weak and fail to bloom properly.

Management :

- Early removal of infected leaves/plants helps to prevent the further spread of the disease.

SEPTORIA LEAF SPOT (SEPOTRIA CHRYSANTHEMELLA)

Leaf spots occur during cool-wet periods of the rainy season:

- Since the pathogens are spread through rain splashes the lowermost leaves get infected first.
- Serious infection may result in premature withering of the leaves; the dead leaves hang to the stem for some time.
- When flowering starts the infection occurs on flower buds, which rot completely.

Management:

- Destruction of disease debris and avoiding excessive irrigation is recommended.

POWDERY MILDEW (OIDIUM CHRYSANTHEMI)

Infection is more severe in older plants under humid conditions:

- The growth of the fungus on the leaves appears as powdery coating. Infected leaves turn yellow and dry out.
- Infected plants remains stunted and fail to flower.

Management

Good ventilation and proper spacing for free circulation of air is recommended.

DISEASES OF CARNATION

FUSARIUM WILT (FUSARIUM OXYSPORUM F.SP. DIANTHI)

Symptoms

In young plants, the first sign of the disease is fading or graying of the normal colour of the leaves with wilting of the leaves and young stems. It is followed by eventual collapse of the whole plant. When older plants are infected, similar symptoms are produced but the older leaves may show chlorosis followed by an indistinct purple-red discolouration. The vascular tissues of infected stems is stained dark brown. Mature plants show wilt symptoms over a period of several months before they die and eventually become straw coloured. The diseased plants should be removed immediately after noticing the disease.

Complete root system and surrounding soil should be dug out and disposed off carefully. Soil solarization using clear transparent polyethylene film (0.1 mm thick) for 30 days gives satisfactory control. Grafting of susceptible cultivars like A lice, Fulvio Rosa, Gus Royalette and Johy, on to resistant rootstocks *i.e.* Arancio 25D, Exquisite, Heidi and May Britt and growing in soil naturally infested with fungus was also found to reduce the incidence of disease.

ALTERNARIA LEAF SPOT (ALTERNARIA DIANTHI)

Symptoms

The chief symptom is blight or rot at leaf bases and around nodes, which are girdled. Spots on leaves are ashy white. The centre of old spots are covered with dark brown to black fungal growth. Leaves may be constricted and twisted and the tip may be killed. Branches die-back at the girdled area and black crusts of conidia are formed on the cankers.

To reduce the disease incidence, humidity may be kept low by providing proper air circulation. Disease-free planting material should be used.

BACTERIAL WILT (BURKHOLDERIA CARYOPHYLLI)

Symptoms

The upper parts of established plants turn pale and wilt.The stem develops elongated discoloured stripes and split open which is characteristic of the disease.The roots are rotted partially and the cortical tissues become sticky and shows discoloration, a tendency to straighten out instead of remaining curled. The leaves are twisted. The roots are generally lacking on one side and remain discolored.

The base of the cutting is discolored, with an elongated brown area extending upwards.The bacteria invade xylem vessels and spread to infect the young shoots. The vessels are disrupted and the host plant produces meristematic tissues. This new tissue causes an uneven development of cortex which splits open longitudinally. Carnation wilt, a vascular disease may be induced by mechanical plugging, a toxin or a combination of factors. Carnation cuttings wilted when placed in a filtrate of a bacterial suspension of B. caryophylli. In the naturally infected carnation plants the xylem vessels are found partly plugged. Lysis of xylem vessels, is also observed. Infection is carried to upper portions of the branch into the leaves. Bacteria are not found in parenchyma tissues. Carnation cuttings suspended in bacterial filtrates in which proteins and other large molecular compounds are removed did not wilting symptoms.

Use of cuttings taken from upper parts of the healthy stock plants are less liable for infection and hence advocated as a control measure. Diseased plant debris should be collected and burnt. Overhead watering and splash watering should be avoided. Disease-free planting materials are to be used. Role cultivars *viz.*, Elegance, Northland and Starlite are less susceptible to bacterial wilt.

DISEASES OF GERBERA (GERBERA ASPLENIFOLIA, G. AURANTIACA, G. JAMESONII (BARBETON DAISY), G. KUNZEANE AND G.VIRIDFOLLA)

ROOT ROT (RHIZOCTONIA SOLANI KUHN AND PYTHIUM IRREGUALSRE)

The infection result in stunted growth. Ultimately the entire plant dry. Rhizoctonia solani causes more losses and can attack older plants. Soil sterilization controls the diseases.

FOOT ROT AND ROOT ROT (PHYTOPHTHORA CRYPTOGEAS)

The short stems blacken and rot. The leaves and flower die. Soil sterilization with vampam at 100 ml/square meter is very effective. Warming of soil (26° C) reduces the incidence of the disease.

BLIGHT / GREY MOULD (BOTRYTIS CINERA PERS)

The fungus kills young growing tissues. The flower heads of the Gerbera growing in humid conditions show small, black spots on the ray florets. Deep planting, poor drainage and poor ventilation predispose the plants to infection. The disease can be reduced when the infected parts are removed and destroyed.

POWDERY MILDEW (ERYSIPHE CHICHORACEARUM DC AND OIDIUM ERYSIPHOIDES F.SP. GERBERA)

The fungus forms white powdery coating on the foliage. Spraying with wettabel sulphur controls E. Chichoracearum. Diseased leaves should be removed and destroyed.

ANTHRACNOSE (COLLETOTRICHUM GLOEOSPORIODES PENZ)

On Gerbera jamesonii Bolus ex. Hook.f. It has been reported from Karnataka and Maharashtra. The disease appears as circular, scattered, reddish brown spots. They coalesce with one another during moist weather involving large area and resulting in withering, rolling and drying of leaves. Excessive watering and crowding of plants should be avoided. Diseased leaves should be collected and burnt.

BLOSSOM BLIGHT ASND STALK ROT (PHYTOPHYTHORA PALMIVORAS BUTLER)

It was reported from Maharashtra during 1971 and Karnataka. The disease appears as light brown, irregular, water soaked spots on flower stalks and petals. The spots increase rapidly and coalesce with one another and form distinct depressed lesions. Under humid conditions the infections become severe involving the entire flower head and resulting in blossom blight and stalk rot. The disease is favoured by drizzling rains and cool moist weather. The fungus is soil-borne and the infection starts from the base touching the soil. Use of disease free soil for cultivation reduces thc disease incidence. Affected flowers should be collected and destroyed. Excessive watering should be avoided

TOBACCO RATTLE (TOBACCO RATTLE VIRUS)

Yellow or black annulated ring spots asre observed on the foliage. Soil steaming before each crop destroys the nematode vector, Trichodorus spp. and prevents the disease.

JASMINE PLANT PROBLEMS: HOW TO TREAT COMMON DISEASES OF JASMINE

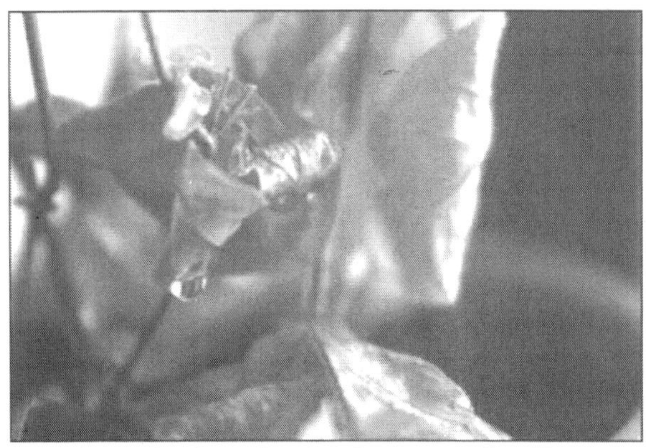

Jasmine flowers bear the intoxicating fragrance familiar to us from perfumes and finely scented toiletries. The plants have an exotic appeal with starry white flowers and shiny leaves. The plants may be grown outdoors orindoors and are fairly easy to grow. However, jasmine plant problems do exist and it is important to be able to identify them. Disease in jasmine plants is easily recognizable and usually the result of cultural issues and easily corrected.

DISEASE IN JASMINE PLANTS

Jasmine plant problems do not occur very frequently, and the plant thrives if it receives attention that mimics its tropical to sub-tropical native region. Jasmine diseases can threaten the foliage, roots, flowers and overall health of the plant. Sometimes they are the result of excess moisture in overly warm conditions; sometimes an insect vector is the cause. The variety of common diseases of jasmine is as broad as it is in any plant, but the first step to diagnosis is to ensure you are giving proper care.

COMMON DISEASES OF JASMINE

Once you are sure the soil type, lighting, heat level, nutrient and moisture acquisition needs of your plant are all being met, it is time to investigate other causes of jasmine plant problems. Foliar problems are common with jasmine plants because they like to live where temperatures are warm and slightly humid. These conditions are most favourable for a variety of fungal diseases.

The most common diseases of jasmine are blight, rust and Fusarium wilt, all of which affect numerous other varieties of plants. These are primarily diseases of the leaves and stems which leave necrotic areas, discolored halos or patches, wilted leaves, streaked stems and occasionally spread to young vegetation. Treating jasmine plant diseases from fungal issues requires a fungicide or baking powder and water spray. Prevention is more crucial because once the fungal spores are active, they are difficult to get rid of. Avoid overhead watering and allow plenty of circulation around the plant to help reduce the chances of fungal issues.

Root knot galls also cause leaves to drop and discolor but mostly this is due to the damage of the nematodes, which are munching away on their roots. The larvae are very difficult to remove once entrenched but you can try a soil drench with an appropriate insecticide. Otherwise, only buy resistant varieties of jasmine.

TREATING JASMINE PLANT DISEASES

Step one in combating any disease is to isolate affected plants.

Step two requires the removal of damaged and diseased plant parts. This includes picking up dropped leaves. In the case of potted plants, installing the jasmine in a clean, sanitized pot with fresh soil will often prevent further fungal spores from damaging the leaves. Then follow proper water and cultural practices to prevent further common diseases in Jasmine.

In ground plants are a bit more difficult, but you can dig around the plant and put in fresh soil or completely remove it and wash off the roots and replant it in a newly amended site. Use Neem oil sprays for any insect issues, fungicides or a mixture of baking soda and water to combat fungi and correct cultural care to promote the health of the plant and help it recover its beauty.

DISEASE CONTROL IN FLOWERS AND SHRUBS

There are many pests and diseases that could cause serious damage to crops and plants in England if allowed to become established. Therefore there are stringent measures that ensure pests and diseases cannot be brought into the country and, if they are, how they must be dealt with.

It is important that you can recognise early signs of ornamental pests and diseases in your plants in order to deal with the problem and - if necessary - inform your local plant health and seeds inspector. This guide explains the symptoms, problems and methods of control for various diseases that affect ornamentals, including P ramorum and P kernoviae. It also covers pests that affect ornamentals, such as nematodes, and the licences and regulations you must follow when importing cut flowers.

STATUTORY ACTION FOR SUDDEN OAK DEATH P RAMORUM AND P KERNOVIAE

Phytophthora ramorum (P ramorum) is an exotic fungus-like plant pathogen which causes damage to trees, shrubs and other plants. It is also known as 'sudden oak death'. In the UK it has been found mainly on container-grown rhododendron, viburnum and camellia plants in nurseries.

It has also been found on various types of:

- Heathers
- Witch-hazel
- Laurel
- Honeysuckle
- Lilac
- Yew
- Californian bay laurel
- Oak species
- Other trees - ash, European beech, horse chestnut, sweet chestnut, sycamore and Winter's bark

Symptoms

Symptoms of P ramorum vary depending on the plant affected. Signs of infection for:

- Rhododendron include blight of shoots, twigs and leaves
- Viburnum include stem base infection or brown to black leaf infection
- Pieris include brown stem lesions leading to aerial dieback, similar to Rhododendron
- Yew include needle blight of young foliage leading to aerial dieback
- Trees include blackening of bark, leaves, or shoots. Bark infections appear as large cankers with discoloured outer bark that seeps red sap. When the outer bark is removed, inner bark has dead inner-bark tissue

The disease spreads through asexual spores produced on leaves of susceptible hosts.

These can survive in plant debris for two winters and are dispersed by:

- Rain splash
- Wind driven rain
- Irrigation
- Ground water

DISEASE CONTROL

Since the first confirmations, there has been a co-ordinated approach to controlling the disease, which aims to:

- Contain the disease
- Eradicate the disease
- Gather evidence to create future policy

Action is only taken when the disease is confirmed and usually includes the:

- Destruction of affected plants
- Tracing of related stocks on horticultural plants moving in trade
- Increased monitoring of imported host plants

P ramorum is a notifiable pest and statutory action is being taken to prevent its introduction and spread. If you believe the pest may be on your premises, you should immediately contact the Fera

Phytophthora kernoviae (P kernoviae) P kernoviae is a fungal disease that causes similar damage and has similar symptoms to P ramorum.

Symptoms

P kernoviae produces similar symptoms to P ramorum and has been found present in the same hosts, though most findings have been on rhododendron.

The main symptoms are:

- Leaf blackening leading to necrotic lesions
- Cankers on tree barks
- Bleeding lesions on tree trunk - often dark blue to black

The infection can spread from leaf to leaf or plant to plant through:

- Water splash
- Airborne mist droplets
- Movement of contaminated plant material

Disease Control

As P kernoviae is a recently described species, it has only been observed in a limited number of sites. Currently, it is being treated as part of the P ramorum study.

FIREBLIGHT

Fireblight is a serious disease caused by the bacterium Erwinia amylovora of apples, pears, Rosaceae trees and shrubs.

Its hosts include:

- June berry
- Flowering quince
- Cotoneaster
- Hawthorn
- Quince
- Loquat
- Apple
- Medlar
- Firethorn
- Pear
- Mountain ash

It is a quarantine disease, to prevent it moving - particularly into designated 'protected zones'. However, it is now widespread throughout South and Central England due to its ability to move in planting material or aerially, eg along hawthorn hedges planted near railways, motorways and main roads.

Symptoms

Fireblight can affect all aerial parts of the host. Symptoms include:

- Wilting and death of flower clusters after a blossom infection
- Withering and death of young shoots - the tip of the shoot can bend to form a 'shepherd's crook' shape
- Leaves with necrotic patches, which spread from the leaf margin or the petiole and midrib, depending on the initial site of infection. These generally remain attached to the plant
- Infected fruit turns brown or black and shrivels - however it usually stays attached to the plant
- Cankering which may spread into the main stem and kill the plant by girdling. Externally, the cankers are usually sunken in appearance and surrounded by abnormal cracks in the bark. When the bark is removed, a reddish-brown discolouration of the underlying tissues may be revealed, often with a well-defined leading edge to the stained area

It spreads through infected planting material which often does not show symptoms. Initial infection occurs through wounds in young shoots or blossom. The disease can then spread through the main stem and kill the plant.

DISEASE CONTROL

There are no chemical measures available to control fireblight. Nurseries selling host plants must register and be approved for issuing plant passports. If an infection is located, the Plant Health and Seeds Inspectorate (PHSI) has statutory powers to require removal of infected plants.

Premises that are registered to issue passports for plants being traded into EUProtected Zones must meet additional requirements including buffer zone freedom. If infection is confirmed on or near registered premises, the PHSI has statutory powers to prevent its spread by removal of the affected plants.

NEMATODES

Nematodes are small, transparent, threadlike worms that are a common pest in the UK.

Most nematode problems in ornamentals are caused by:

- Aphelenchoides ritzemabosi (A ritzemabosi) and Aphelenchoides fragariae - leaf and bud nematodes often found on nursery stock and herbaceous plants
- Ditylenchus dipsaci (D dipsaci) - stem or bulb eelworm often found on bulbs and herbaceous plants
- Meloidogyne spp - root-knot eelworm

As well as damaging plants through feeding, nematodes can also transmit viruses.

Nematode Symptoms and Damage

A ritzemabosi attacks plant parts above the ground and can show a number of symptoms including:

- Bushy plant appearance - first stems are dwarfed and the plant must grow extra basal stems
- Distorted and deformed leaves - feeding in the buds causes growth retardation
- Blackened or dead leaves
- Infected side shoots
- Infection of leaves leading to yellow spots and blotches - a later symptom
- Water-soaked bands and blotches on ferns - leaves become dark brown or black and eventually die

Symptoms or damage from D dipsaci include:

- Discolouration or distortion of tissues - feeding of nematodes may kill affected plant tissues or the entire plant
- Infested bulbs that are soft at the neck and show discoloured brown rings of dead tissue when cut through
- Misshapen plant growth with distorted leaves or flowers

Stem nematodes can be spread in bulbs, bulb fragments or leaf debris through:

- Wind blown contamination
- Cultivation
- Movement of surface water
- Presence in the soil

Meloidogyne spp symptoms are often the presence of 2-centimetre long galls on roots, which can be an irregular shape. Nematodes attacks disrupt normal root function and plants will show nutrient deficiency, leading to wilting and death. Nematodes can only be confirmed by laboratory examination.

TREATMENTS

There are several treatments you can consider to prevent nematode infestation. Biological control - 'green manure' crops can control nematodes if you use them to clean up land before planting. The most effective crops are members of the Brassica family which are effective when chopped and introduced into soil. Bio-insecticides - French marigold can be used to treat land between crops of field-grown trees and herbaceous plants. The variety 'Ground Control' is used in Dutch lily production and a kill rate of up to 95 per cent has been claimed. Plant on land without crops or weeds to ensure marigold roots are the only nematode food source. A biocide marigolds contain can kill the pests. Cultural control - good general weed control will prevent the establishment and spread of stem and bud eelworms.

You should consider using hot water treatment to kill any infestations in herbaceous perennials. You should only expose plants to hot water at 44.4°C for 15 minutes to avoid damaging plant tissues. Bulb treatment requires a longer period of time. You can also reduce the risk of flower damage by storing bulbs at 30°C for three weeks before treatment. Leaf and bud nematodes can be transferred in cuttings and spread on propagation tools, dead and dying plant material and poorly composted material. You can avoid excessive moisture in propagation and space plants to avoid leaf contact. You should use disinfectants to routinely clean tools, equipment and surfaces.

PREVENTION

You can follow similar prevention guidelines for all nematode species. You should ensure that:

- Narcissus bulbs are sourced from a certified crop wherever possible and must be accompanied by a plant passport which confirms bulbs are free from stem nematodes. There is currently no certification scheme for tulips but they must have a plant passport. All passport plant stocks are inspected in the growing season by the Defra Plant Health and Seeds Inspectorate
- Crops susceptible to attack by stem nematode should only be grown in the same field every four years. Groundkeeper bulbs should be destroyed
- Machinery used during bulb production is regularly cleaned and disinfected. Bulb waste from grading should not be returned to agricultural land

- Growing crops are regularly inspected for signs of stem nematode damage. Suspect bulbs and the bulbs next to them should be destroyed (destroy bulbs within a 1 metre radius - nematodes can move up to 1 metre in a season)
- All narcissus stocks are hot-water treated
- Infected plants and fallen leaves are carefully removed from the garden or greenhouse and burned. Soil in affected nursery plantings should be fumigated

PEST AND DISEASE THREAT WHEN IMPORTING ORNAMENTALS AND CUT FLOWERS

You must follow certain plant health controls when importing cut flowers into the UK to prevent the introduction of plant pests and diseases.

For plant health purposes the three categories plant materials fall into are:

- Prohibited - Poses such a serious risk that import is only permitted under authority of a licence issued by the Fera or the Forestry Commission
- Controlled - This includes cuttings, rooted plants and trees that are not prohibited, as well as bulbs, most fruits, certain seeds and some cut flowers. You usually need a phytosanitary certificate from the plant protection service of the exporting country
- Unrestricted - Presents little or no risk and is not subject to routine plant health controls. Includes nearly all flower seeds, some cut flowers and fruit, and most vegetables for consumption or processing - except potatoes

Licences

You can sometimes import prohibited or controlled plants under licence from Fera. You must make this application before you import and you may have to pay a fee.

Feracan issue licences for the import, movement and keeping of:

- Plants and plant material for scientific or trialling purposes
- Soil and growing medium for physical or chemical analysis

Imports of certain forest trees, wood, bark and some wood products are subject to legislation implemented by the Forestry Commission.

COMMON FLOWER DISEASES

BATTLING MILDEW, MOLD AND BLACK SPOT

Mid- to late summer is when most plant diseases start becoming noticeable. Powdery mildew, gray mold (Botrytis), and black spot are three of the most common flower diseases. Learn how to recognize them and what you can do to avoid them in your own garden.

Powdery Mildew

Powdery mildew, as the name suggests, resembles a white, powdery coating on leaf surfaces. If severe, it also might appear on stems and the flowers themselves. Affected leaves eventually turn yellow, then brown. Dead foliage typically falls off the stem, though it will sometimes remain in place. Although not fatal to plants, powdery mildew makes the foliage unattractive and repeated bouts of the disease will gradually weaken the plant. Annual flowers that are particularly susceptible to powdery mildew include zinnias, snapdragons and verbena. Perennials that are commonly infected include delphiniums, lungwort, bee balm and garden phlox.

Prevent Disease with Proper Watering

As with most fungal diseases, gray mold is spread by wind and splashing water.

To prevent problems, avoid getting irrigation water from splashing onto the leaves and flowers:

- Instead of sprinklers and spray wands, water gently at the soil level, near the base of the plant. Or, better yet, use drip irrigation.
- Make a habit of watering in the morning, rather than the evening, so foliage has time to dry out during daylight hours. In most cases, disease spores can only infect foliage if it is consistently moist for several hours.

Most fungal diseases are spread by microscopic structures called "spores" that are transfered on wet foliage. However, powdery mildew thrives in high humidity. Keeping plants well-spaced and removing weeds will help ensure good air circulation and reduce the humidity around plants.

Control and Prevention:

- Choose varieties that have proven to be resistant.
- Use a horticultural oil, insecticidal soap or another spray. As with all fungus diseases, it is essential to begin application at the early onset of the disease— often late June or early July — and ensure that all susceptible foliage is treated. Repeated applications are usually necessary right through the duration of the growing season. Read and follow the instructions on the label. Options include:
  - Neem oil, derived from the neem tree.
  - Serenade Garden Disease Control is a spray that contains beneficial bacteria.
  - Garden Dust, an insecticide-pesticide that contains copper.
  - Try a home remedy with baking soda: Mix 1 teaspoon per quart of warm water; spray on plants every seven to 10 days.
- Ensure good airflow around susceptible plants; don't crowd them together.

Gray Mold

Gray mold is perhaps the most common disease of flowers. It is especially problematic during periods of high rainfall and cool temperatures. Like powdery mildew, gray mold is well named. It appears as a gray mold, primarily on old and dying leaves and flowers.

It begins as water-soaked spots and eventually develops into the characteristic gray, fuzzy coating:

- Remove diseased flowers and/or leaves right away.
- Spray with beneficial bacteria, such as those found in Serenade Garden Disease Control.
- Thin plants and keep flower beds weeded to allow plenty of air circulation around plants.
- Avoid overhead watering.

Black Spot

Black spot is another common fungal disease. It is a big problem with roses. The disease typically begins as black spots on the foliage. These spots are most prevalent on upper leaf surfaces, and may be up to ½3 across. Leaves eventually begin to yellow around the spots, then become all yellow and fall off. The spots may also appear on rose canes, first being purple and then turning black.

Control and Prevention:

- Black spot requires at least seven hours of wet conditions for infection, and it is inhibited at temperatures about 85 degrees F. Although you may not be able to turn up the temperature in your garden, you can minimize the disease by keeping foliage dry through proper watering and good air circulation.
- Plant susceptible flowers, such as roses, in an open and sunny location and avoid watering during cloudy weather.
- Black spot fungus overwinters in fallen leaves and infected canes. Prune out infections and rake the fallen leaves at the end of the season.
- Rose varieties vary greatly in their resistance to black spot, so choose resistant ones. If this information is not indicated on plant labels, you can research the variety online or check with experts at your local nursery.
- Several of the shrub roses show resistance to black spot—and powdery mildew. Look for cultivars from the Meidiland, David Austin and Explorer series.
- Many of the other shrub roses and old-fashioned roses are resistant as well.
- Neem oil can help prevent black spot. Spray every 10 to 15 days during the growing season.

DOWNY MILDEW OF FIELD PEAS

Downy mildew, caused by the pathogen Perenospora viciae, is a common disease of peas in Victoria, South Australia and Tasmania. Downy mildew is one of the most common fungal diseases of peas, often causing a substantial reduction in plant numbers in wet, cool seasons.

The disease also impairs wax formation on the leaves and makes plants very susceptible to herbicide damage. Substantial losses are likely to occur in cooler districts.

Fig. Thick Grey Fungal Growth on Lower Leaf Surface, this is Typical of Downy Mildew.

Fig. Upper Leaf Surface Turns Yellow Above Fungal Growth.

WHAT TO LOOK FOR

The disease is most common soon after emergence but may affect plants at any growth stage, during periods of moist, cool weather. Systemically infected plants are a sickly yellowish-green and severely stunted and distorted. The undersides of the leaflets, in particular, are covered with a fluffy mouse-grey spore mass. Infected plants may turn yellow while producing an abundant source of spores for secondary infections.

Secondary infection results in the appearance of isolated greenish-yellow to brown blotches on the upper leaf surface, while directly under these lesions on the lower surface are masses of mouse-grey fruiting bodies. The fungus usually affects the lowest leaves and pods. Infected pods are deformed and are covered with yellow to brownish areas and superficial blistering.

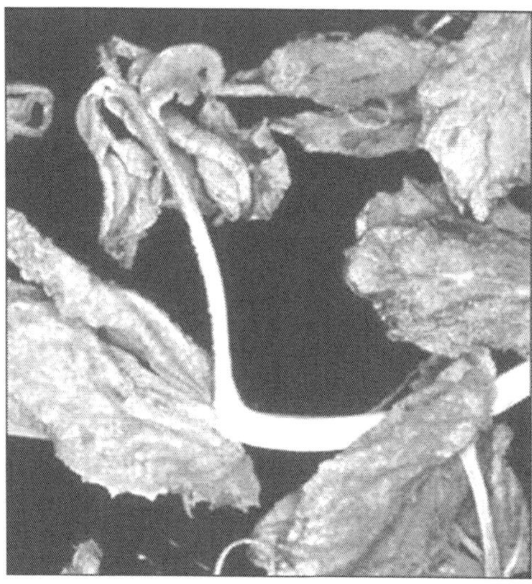

Fig. Downy Mildew on Older Plants.

DISEASE CYCLE

The fungus that causes downy mildew survives in the soil and on pea trash and can be seed-borne. Infected seed can act as a primary source for systemic and local infections.

The disease can develop quickly when conditions are cold (5 Ã,– 15°C) and wet for 4 Ã,– 5 days, often when seedlings are in the early vegetative stage. Individual seedlings become infected and act as foci of infection from which the disease spreads.

Rain is the major means of spore dispersal and infection. Heavy dews will promote sporulation. Dry, warm weather is unfavourable for the disease. Systemic infection of plants can lead to the disease developing late in the season if conditions are favourable.

This disease is most damaging by stunting plants early in their growth or death of seedlings in more extreme instances. Generally plants will grow away from the disease as temperatures increase in late winter/early spring without significant yield loss.

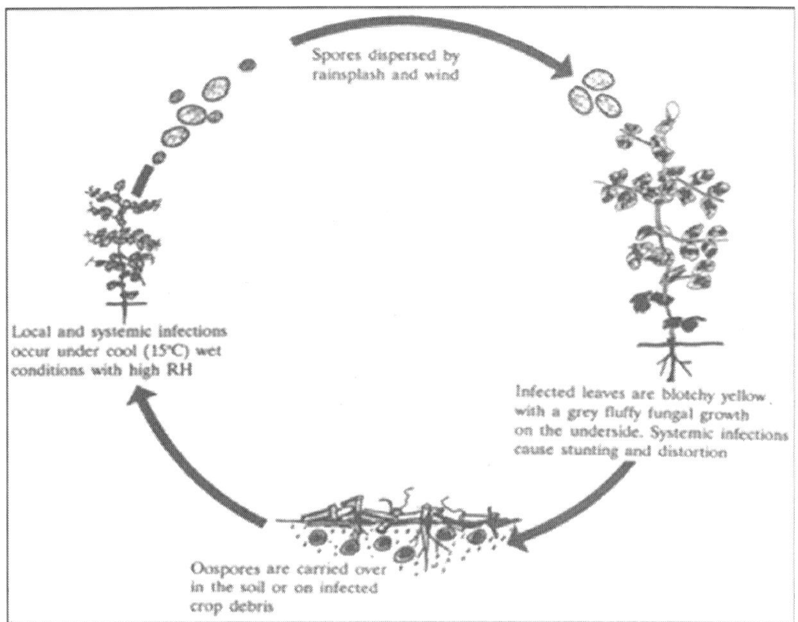

Fig. Disease Cycle of Downy Mildew on Field Peas.

MANAGEMENT

Varietal Selection

Growing a resistant variety is the most effective means of controlling downy mildew in districts prone to this disease. However, there are now two strains of the downy mildew fungus. They are the Parafield strain and the Kaspa strain. The Kaspa strain is a new strain, and prior to its emergence there were commercial varieties available with resistance to downy mildew.

Now with the Kaspa strain detected in Australia, there are no commercial varieties with resistance to both strains of the fungus.

Chemical Control

The main effects of downy mildew can be reduced by treating seeds with fungicides. Seed treatments reduce the number of seedlings with primary infection, thereby reducing the amount of air-borne spores that cause secondary infection in the surrounding crop. Seed treatments can be beneficial, and are recommended for districts where downy mildew occurs in most years. Not all fungicide seed dressings have activity against downy mildew.

Crop Rotation

Extended crop rotations and destruction of infected pea trash will minimise the risk of serious disease. Extended crop rotations allow spore numbers in the soil to decline before sowing again to field peas. A break of at least 3 years between field pea crops is recommended. Avoid sowing pea crops adjacent to last season's stubble.

SEPTORIA BLIGHT OF FIELD PEAS

This disease is caused by the fungus Septoria pisi. Septoria blotch is a minor disease and appears to have little effect on the yield of most pea varieties. The disease has caused severe losses in canning pea crops in the United States.

WHAT TO LOOK FOR

The disease is found mainly on the lower, senescing parts of the plant and the pods. The diseases areas vary in size, are roughly circular, and have no distinct margin. First they are yellow, later becoming straw-coloured. Several such blotches may join to cover the entire leaf.

Within the lesions, pin-point sized black fruiting bodies of the fungus may be seen. Diseased tissues may dry of prematurely.

Fig. Typical Leaf Lesions Caused by Septoria infection. These are Light Brown and Angular, Containing Very Small Brown to Black Spots.

Fig. Septoria can Infect Pea Leaves and Tendrils.

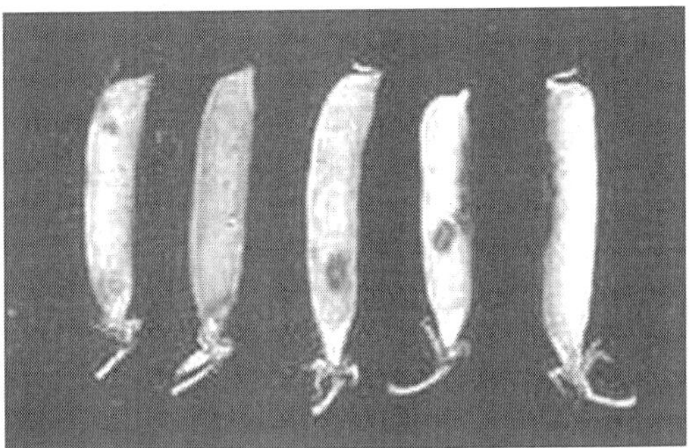

Fig. Late Spring Rainfall will Result in Pod Infection by Septoria.

DISEASE CYCLE

The fungus over winters on infected pea trash and seed. Spores of the fungus are carried, by wind from infected trash, into the new crop. Infection is found on the lower foliage where the humidity is high following rain or heavy dews. Disease is favoured by warm temperatures and extended periods of high humidity. Splattering of water assists in spreading the disease within a field. Whilst seed transmission can occur it is less important.

Management

Septoria can be managed by using an integrated approach that encompasses crop rotation, stubble management and fungicides.

Crop Rotation

The Septoria blotch fungus survives in soil and on old pea trash. It is only safe to re-crop an area with peas after all pea debris has decomposed. Destroying pea stubble by grazing, burning and cultivation will help in reducing the pea debris more quickly. This also minimises the amount of inoculum available to infect new crops.

8

Flowering Plant

INTRODUCTION

Flowering plants are either classified as short- or long-day flowering. While a number of factors influence a plant's florescence, the length of dark photoperiods—also referred to as nights—trigger flowering. Short-day plants flower when the night is longer than a critical length.

These plants cannot flower under long days or if artificial light is shone on the plant interrupting the "night." Short-day plants require a consolidated period of darkness before floral development can begin.

Natural nighttime light is not bright enough nor does it endure long enough to interrupt flowering. While the length of the dark period required to induce flowering differs among species and varieties varietals, short-day plants generally flower as days grow shorter. This flowering period coincides with the time after the summer solstice in the Northern Hemisphere, which is during summer or fall.

Photoperiod affects flowering when the shoot is induced to produce floral buds instead of leaves and lateral buds. Note that some species must pass through a "juvenile" period during which they cannot be induced to flower. Common cocklebur is an example of a plant species with a remarkably short period of juvenility, and plants can be induced to flower when quite small. Long-day plants require shorter periods of darkness than short-day plants to induce flowering. These plants typically flower in the Northern Hemisphere in late spring or early summer.

Day-neutral plants, such as cucumbers, roses and tomatoes, do not initiate flowering based on photoperiodism at all; they flower regardless of the night length. They may initiate flowering after attaining a certain overall developmental stage or age or in response to alternative environmental stimuli, such as vernalisation (a period of low temperature), rather than in response to photoperiod.

Photoperiodism explains why some plant species can be grown only in a certain latitude. Experiments with the cocklebur have shown that the term "short-day" is a misnomer. In order to flower, cocklebsurs need a long dark photoperiod (of at least 8.5 hours).

Interruption of the dark photoperiod prevents flowering unless it is followed by irradiation with far red (730 nm) light. Exposure to far-red light at the start of the night reduces the dark requirement by 2 hours.

While commercial growers have had success in altering photoperiods to force plants to flower, for many plants, flowering is much more complex than this. Some require the correct photoperiod for several days before they will respond. Additionally, some not only require the correct photoperiod but also other environmental cues.

PHOTOPERIODISM

The reproductive cycles of many organisms, both plant and animal, are regulated by the length of the light and dark period, called the photoperiod. In flowering plants (angiosperms), flowers are organs for sexual reproduction, and photoperiodism refers to the process by which these plants flower in response to the relative lengths of day and night. Along the equator, the lengths of day and night remain constant because the sun rises and sets at the same time throughout the year.

The lengths of the day and night are also equal (each is six months long) at the exact North and South Poles due to the fact that the sun remains below the horizon for six months each year and above the horizon for the other six months. Any place else in the world, the days become longer in the summer and shorter in the winter.

The reproductive cycles of many organisms, both plant and animal, are regulated by the length of the light and dark period, called the photoperiod. In flowering plants (angiosperms), flowers are organs for sexual reproduction, and photoperiodism refers to the process by which these plants flower in response to the relative lengths of day and night.

The synchronization of reproduction with seasonal time is a very important aspect of plant physiology.

Reproduction in many angiosperms is dependent on cross-pollination, the process of pollen being transferred from one flower to another. Hence, it is important for all of the plants of the same species in a given region to flower at the same time.

STRUCTURE OF PLANT

The points on the stem to which the leaves are attached are usually slightly thickened, and are called nodes or 'joints' the lengths of stem between them being termed internodes. It has been already noticed that the seedling bean plant consists of a descending portion, the root, and an ascending part which comes above ground.

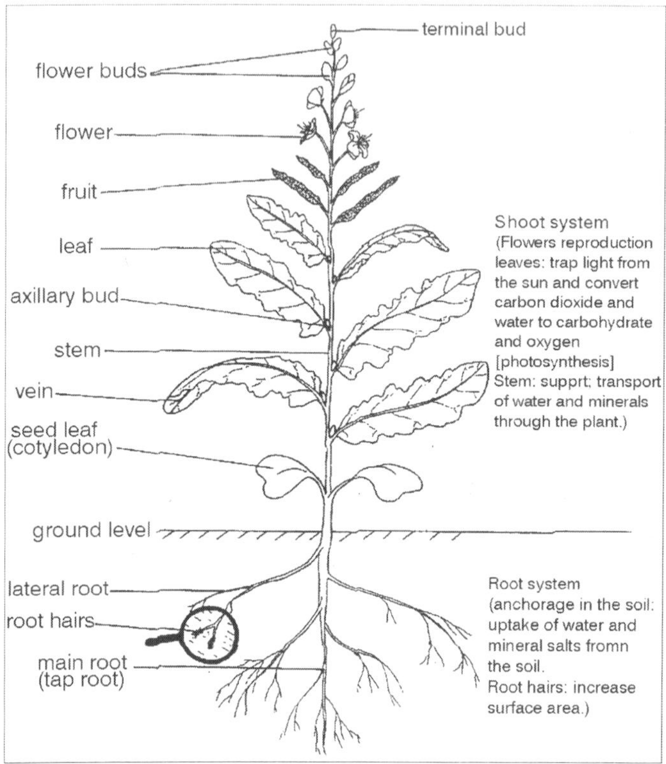

Fig. Structure of Plant

The latter is known as the primary shoot, and consists of an axis—the stem—upon which are arranged a series of lateral appendages which are called leaves. Flowers ultimately arise upon the shoot, and it is one of the characteristics of seed-bearing plants that seeds are always produced on their shoots and never on roots.

For the present, however, flowers may be left for future consideration, and attention paid to the origin and nature of the vegetative shoot or stem with its ordinary green leaves. In the earliest stages of the development of a bean plant the primary shoot is very short and bears the cotyledons or primary leaves, its tip ending in the plumule.

The latter is a bud, and at the time when the seed commences to germinate, its parts cannot be fully made out by observations with the naked eye. As soon as it comes above ground, however, the bud is found, on examination, to consist of a short stem hidden by a number of enfolding leaves. As growth proceeds, the short stem inside the bud elongates, and the leaves, which at first are crowded upon it, become separated from each other.

Marks made upon the stem as previously explained for the root, reveal the fact that the increase in length takes place at the tip of the shoot. After reaching a certain length the lowest intervals between the leaves cease to elongate; the upper, younger and shorter ones also lengthen and cease in a similar manner, to be followed in turn by still younger parts of the stem nearer the tip.

The stem, before the growing season is over, may thus reach a height of two or three feet, or even more, the extreme tip, or growing-point as it is called, remaining young all the time, and acting as a manufactory for the addition of more stem and leaves. The growing-point, which is of a tender and delicate nature, is protected by the enfolding young leaves, the latter arising as outgrowths from its external surface.

The youngest leaves are always nearest the tip the stem which bears them, the older ones being further removed from it in regular order—that is, they arise in acropetal succession, and adventitious leaves are never met with. Instead of the latter growing at once into a long shoot, bearing leaves at some distance from each other, the primary axis within the plumule elongates very little, its internodes remain very short, and the leaves arising upon it appear crowded together, usually in the form of a rosette, a short distance above where the cotyledons were placed; this form of stem with short contracted inter-nodes is well illustrated in the first season's growth of mangels, turnips, carrots, certain thistles, and red clover.

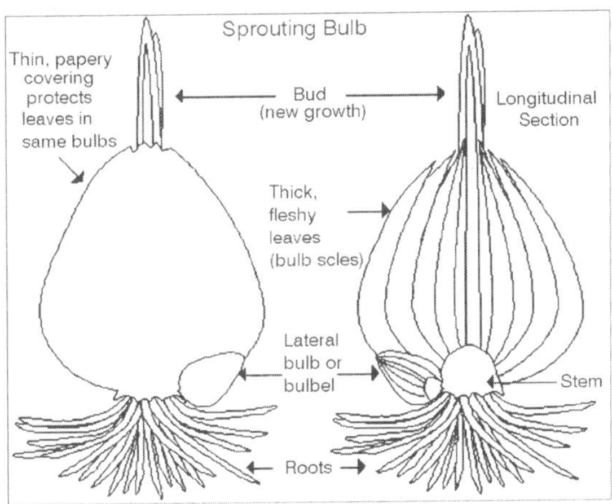

Fig. Growing Stem of Onion

In such plants as these, the primary root and hypocotyl become much thickened by the deposition within them of reserve-food prepared by the leaves, and it is only during the following year that the growing-point of the stem, which is hidden in the centre of the rosette, elongates and produces a shoot with long internodes, and bearing a series of new leaves at considerable intervals.

In the onion and many bulbous plants the primary stem also remains very short, and the reserve-foods prepared by them are deposited in the bases of the leaves, instead of being stored in the root and stem, as in the former instances.

BUDS

The stems and leaves of all flowering plants originate from buds in the manner indicated above; buds may therefore be termed embryonic or incipient shoots. It is by their growth that trees, which appear so bare in winter, become clothed with fresh green leaves in the succeeding spring. The relationship which they bear to the leaves and stems produced by them is easily discerned by examining the structure and watching the development of the terminal bud of a young sycamore tree.

On the outside is observed a series of scaly leaves, which overlap each other, and protect and cover up the delicate growing point of the twig. A section through the bud shows the disposition of these scaly leaves and within are also seen the ordinary leaves arranged upon a very short stem (s). In spring the inner scaly leaves grow for a time.

The stem which bears the rudimentary green foliage-leaves, elongates, and the latter are pushed out from between the protective scaly leaves of the bud. After a week or ten days, the stem has reached a considerable length, and the leaves, which were rudimentary and packed away in the bud, unfold themselves and grow out flat. The number of foliage-leaves present upon a developed shoot is often indicated in the bud, but in some plants, especially those of a herbaceous character, the growing point of the bud continues to produce new leaves until frost checks it in autumn.

The vegetative shoots of plants usually end in terminal buds and an examination of almost any kind of plant shows that not only are buds present at the tips of the stems, but on their sides as well. These lateral buds arise ordinarily in the upper angles formed where the leaf-bases and stem join each other. The angles are termed the axils of the leases, and the buds are designated axillary buds.

Most frequently only one bud is produced in each leaf-axil, but in some instances, two or more may be present. Generally the first leaves of the bud which are outermost or lowest down on the stem, are rudimentary structures, smaller and different in appearance from those which unfold later.

In the primary bud or plumule of the bean, and many similar herbaceous plants, this is observable, but it is most evident in buds which are met with upon perennials, such as shrubs and trees. In the latter the outermost leaves of the buds are generally more or less firm, tough structures, termed scales or scale-leaves, which protect the interior of the bud from being injured by frost, rain, and other agents during the winter. Buds, such as those of the sycamore and pear, having scales are termed scaly buds, those without, such as mealy guelder-rose, being known as naked buds. Buds similar to those of the bean and sycamore, previously described, which develop into shoots bearing green foliage-leaves, are termed leaf-buds: when met with upon trees they are sometimes named wood-buds, as it is from them that new woody twigs are produced.

Many buds, however, on opening, give rise to flowers only, and are termed flower-buds: a third kind is met with producing short shoots bearing both green leaves and flowers; these are mixed-buds.

Among gardeners the two latter forms are known as fruit-buds as it is from them that fruit is obtained. It is not possible in all cases to distinguish fruit-buds and wood-buds by their outward appearance, although for budding and pruning operations and the general management of fruit-trees it is desirable to do so. In apples and pears the wood-buds are small and pointed, the fruit-buds being blunter, more plump, and of larger size.

In cherries and plums both kinds are very similar in appearance, and it is only in spring when they begin to develop that the stouter and blunter characters of the fruit-buds show themselves. Their position upon the shoots is a great aid in distinguishing the two classes of buds.

CHARACTERITIC OF FLOWERING PLANTS

Plants, to begin with go through a period of vegetative growth. The extent of vegetative growth is endowed with its genetic potentiality. Accordingly, they may grow into herbs or shrubs and some may develop into trees or climbers.

Generally, every plant after going through a period of vegetative growth, responding to environmental clues, start producing floral structures, which may be in the form of characteristic single flowers or inflorescences.

Many plants for that matter, a large number of plant species (higher plants), after a period of vegetative growth, start flowering irrespective of the season. But some plants flower only in a particular season of the year. Based on the duration required for the plants to produce flowers, they have been classified into annuals, biennials and perennials. All plants have to acquire ripeness to flowering. Annuals complete their vegetative growth and flowering in one season and then they die. Biennials produce vegetative growth in one season and flower in the next season. But perennials remain for many years and flower seasonally.

In fact, some trees do not flower till they reach a certain age. For example, coconut and areca nut plants start producing flowers only when they reach an age of 6-8 years. On the other hand in the case of bamboo plants, they grow for a number of years, and flower only once in their life span. As soon as they flower, produce seeds and plants die (monocarpic plants). Interestingly there are many plants which flower throughout the year, ex., Catharathus roseus.

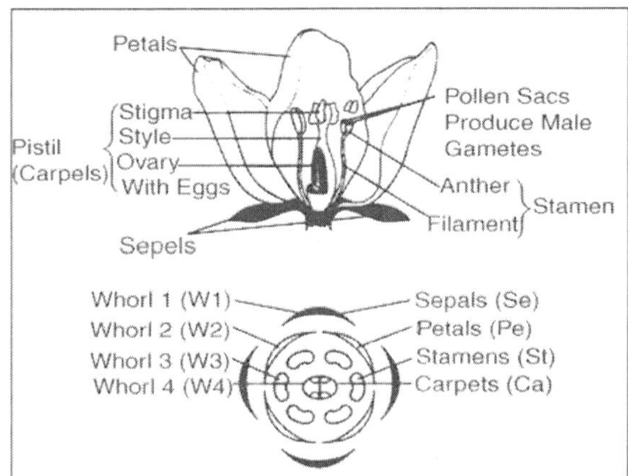

Parts of a flower from outer to the central region

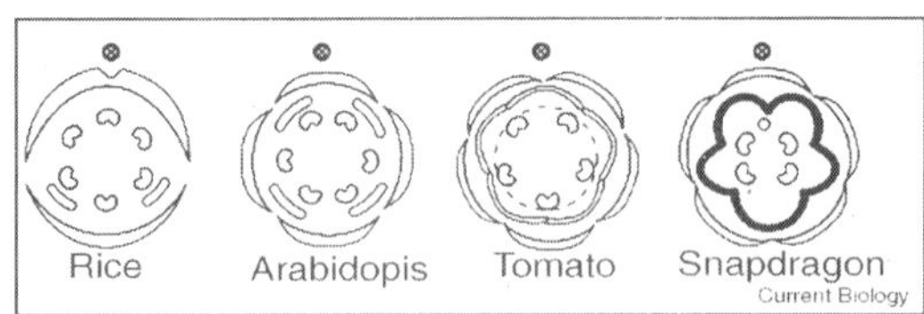

Aestivation of some flowers- refers to the organization of each floral part with respect to each other, especially holds good for sepals and petals.

Plants growing in different regions of the globe are exposed to different climatic conditions and different day lengths. In fact they are adapted to such environs in such a way, they exhibit alternate vegetative and flowering cycles. It means that plants with their inherent genetic potentiality interact with environmental conditions, accordingly, they respond and behave. Humans, homo sapiens present day species, just about 25000yrs to 40000yrs old, copulated with Homo eructus, mostly in Asia and made them extinct; when they evolved and colonized sites of their own, after observing for many centuries the above said natural phenomenon, they have devised different methods to cultivate crop plants in different seasons of the year, so as to get the harvest at the right time of the year. They also domesticated animals for their use.

The common knowledge of the farmer has been extended and explained by plant physiologists; why and how the said plants behave in response to different environmental conditions.

Plants have all the needed signal transduction pathways to respond to environmental signals. Such signal pathways has been worked out in Arabidopsis.

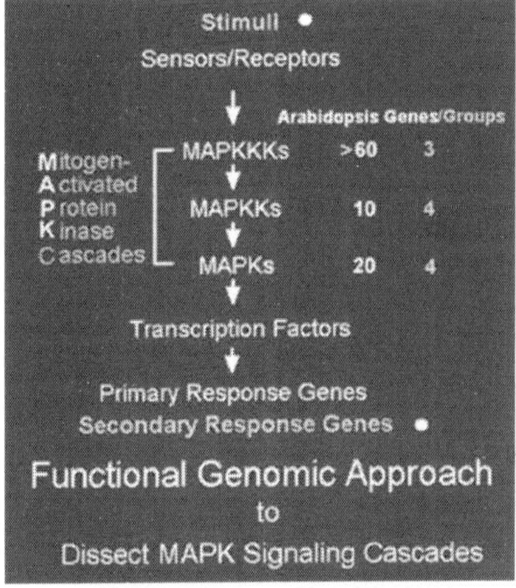

Developmental pathway of plants and its structures start from the zygote and end in all the structures.

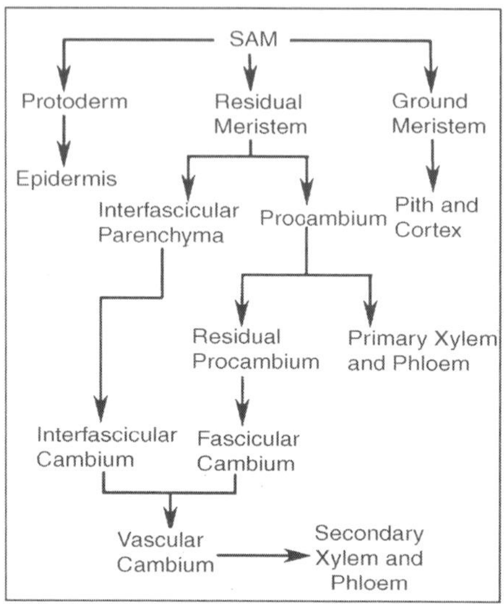

DISCOVERY OF FLOWERING RESPONSE

Plants respond to the changing seasons to initiate developmental programmes precisely at particular times of year. Flowering is the best characterized of these seasonal responses, and in temperate climates it often occurs in spring.

Though it is a common knowledge that different kinds of plants respond to different seasons of the year in producing the flower, it was left to G.Gassner and W.W. Garner to explain the phenomenon by their pioneering scientific studies. Gassner observed that winter variety of petkus rye plants called Secale cereal, responded favourably to cold treatments. Almost at the same period of time, Garner and Allard demonstrated how plants produce flower in response to different lengths of the day and night in a 24 hours day cycle.

The above two phenomenon are popularly called as Vernalization and Photoperiodism respectively. The above studies have lead to the discovery of how plants rhythmically respond and behave to day and night duration or to temperature fluctuation in different seasons of the year and they also observed rhythmical behaviour of the plants which is referred to as 'biological rhythm. And the operational time measuring system found with in the plant structures is called 'Biological Clock.

PHOTOPERIODISM (PP)

Photoperiodism is the physiological reaction of organisms to the length of day or night. It occurs in plants and animals. Photoperiodism can also be defined as the developmental responses of plants to the relative lengths of the light and dark periods. Here it should be emphasized that photoperiodic effects relate directly to the timing of both the light and dark periods.

Many flowering plants (angiosperms) use a photoreceptor protein, such as phytochrome or cryptochrome, to sense seasonal changes in night length, or photoperiod, which they take as signals to flower. In a further subdivision, obligate photoperiodic plants absolutely require a long or short enough night before flowering, whereas facultative photoperiodic plants are more likely to flower under the appropriate light conditions, but will eventually flower regardless of night length.

In 1920, W. W. Garner and H. A. Allard published their discoveries on photoperiodism and felt it was the length of daylight was critical, but it was later discovered that the length of the night was the controlling factor. Photoperiodic flowering plants are classified as long-day plants or short-day plants, even though night is the critical factor, because of the initial misunderstanding about daylight being the controlling factor. Each plant has a different length critical photoperiod, or critical night length.

Modern biologists believe that it is the coincidence of the active forms of phytochrome or cryptochrome, created by light during the daytime, with the rhythms of the circadian clockthat allows plants to measure the length of the night. Other than flowering, photoperiodism in plants includes the growth of stems or roots during certain seasons, or the loss of leaves. Using artificial lighting to induce both extra-long days and nights, short-day plants will bloom but long-day plants will not.

Earth, because of its revolution on its own axis and rotation around the sun; exhibits a period of day and night and seasonal changes. The duration of the day and night again shows variations because of the angle and distance between the earth and the sun at any given time of the year. Thus plants and animals living on different parts of latitudes or longitudes are subjected to different periods of photo periods and different temperatures at different seasons of the year.

If we use three points or places on the globe, located at different positions as the reference point, to measure the day and night periods, it will be apparent how different are the day periods and temperatures of such places. Brazil in South America and Congo in Africa exhibit almost 12 hours of day and 12 hours of night in all the months of a year. But a city like Philadelphia located in the east coast of USA at latitude of 40 degree N, in the month of December; it experiences 9 hours of day and 15 hours of night. On the other hand, in the month of June, the day period is 14 ½ hours and night is 9 ½ hours long. Similarly, cities of Norway, during December, experience 6 hours of day and 18 hours of night, but in June, it enjoys 18 hours of day and 6 hours of night. Such day periods also accompany with changes in extreme temperatures. The above observations suggest that organisms living in these regions are subjected to seasonal variations of day and night and also to changes in seasonal temperature fluctuations.

Garner and Allard, while working the department of Agricultural Station, Beltsville, Maryland, USA, demonstrated remarkable relationship between the effect of the day period and flowering in a mutant tobacco plant called Maryland Mammoth. They observed that the mutant failed to produce flowers but tall, so they are called Maryland Mammoth. They also observed that the same plant started flowering in summer under field conditions.

But the same plant started flowering when transferred to green house where it was subjected to short day and long night conditions. So the plant was called short day plant. Since then, a large number of plants have been subjected to various cycles of photoperiod *i.e.* treatment and according to their responses, plants have been classified into different groups. The flowering response in plants to photoperiodic treatment is now called photoperiodism. Light induced responses in photo morphogenesis are many and intricate, and this can be only represented in the form of network.

CLASSIFICATION

Based on the responses to different photoperiods, most of the plants are grouped into 3 major classes, *viz.* Short day plants, long day plants and day neutral plants. However, detailed studies on each of these groups resulted in further classification of them into sub groups like long short day plants, short long day plants etc. Each of these groups has been further grouped into qualitative and quantitative varieties based on the specificity of the appropriate light periods.

CRITICAL DAY PERIOD

It is the duration of the photoperiod or the dark period that ultimately determines whether the plant has to go through vegetative growth or to produce flowers. Different plants require different periods of light or dark for 100 per cent flowering. If that period falls short then plants do not produce 100 per cent flowering.

Such requirement of a minimum of photoperiod or dark period for effective flowering is called critical day period. The length of light and dark period for different long day and short day plants varies. For example Xanthium requires a critical length of 15 ½ hours of dark period for its effective flowering. If the dark period is less than 15 ½ hours plants do not induce any flowering, but longer dark periods d not inhibit flowering. On the other hand, the long day plant, ex. Hyoscyamus niger requires a critical 11 hours of exposure to light. Anything less than that, plants fail to produce flowers. If the length of the day period for this plant is more than 11 hours, it does not affect the flowering. Similarly, different plants have different critical day periods and the correct photoperiod has to be determined individually by subjecting them to photoperiodic treatment.

LONG-DAY PLANTS

Long-day plants flower when the day length exceeds their critical photoperiod. These plants typically flower in the northern hemisphere during late spring or early summer as days are getting longer.

In the northern hemisphere, the longest day of the year is on or about 21 June (solstice). After that date, days grow shorter (*i.e.* nights grow longer) until 21 December (solstice). This situation is reversed in the southern hemisphere (*i.e.* longest day is 21 December and shortest day is 21 June). In some parts of the world, however, "winter" or "summer" might refer to rainy versus dry seasons, respectively, rather than the coolest or warmest time of year.

Some long-day obligate plants are.

- Carnation (Dianthus)
- Henbane (Hyoscyamus)
- Oat (Avena)
- Ryegrass (Lolium)
- Clover (Trifolium)
- Bellflower (Campanula carpatica)

Some long-day facultative plants are.

- Pea (Pisum sativum)
- Barley (Hordeum vulgare)
- Lettuce (Lactuca sativa)
- Wheat (Triticum aestivum, spring wheat cultivars)
- Turnip (Brassica rapa)
- Arabidopsis thaliana (model organism)

SHORT-DAY PLANTS

Short-day plants flower when the day lengths are less than their critical photoperiod. They cannot flower under long days or if a pulse of artificial light is shone on the plant for several minutes during the middle of the night; they require a consolidated period of darkness before floral development can begin. Natural nighttime light, such as moonlight or lightning, is not of sufficient brightness or duration to interrupt flowering.

In general, short-day (*i.e.* long-night) plants flower as days grow shorter (and nights grow longer) after 21 June in the northern hemisphere, which is during summer or fall. The length of the dark period required to induce flowering differs among species and varieties of a species.

Photoperiod affects flowering when the shoot is induced to produce floral buds instead of leaves and lateral buds. Note that some species must pass through a "juvenile" period during which they cannot be induced to flower—common cocklebur is an example of a plant species with a remarkably short period of juvenility and plants can be induced to flower when quite small.

Some short-day obligate plants are.

- Chrysanthemum
- Coffee
- Poinsettia
- Strawberry
- Tobacco, var. Maryland Mammoth
- Common duckweed, (Lemna minor)
- Cocklebur (Xanthium)
- Maize – tropical cultivars only

Some short-day facultative plants are.

- Hemp (Cannabis)
- Cotton (Gossypium)
- Rice
- Sugar cane

RIPENESS TO FLOWERING AND SITE OF PERCEPTION

Not all photoperiodic plants respond to the light treatment until and unless the plant has grown to certain vegetative maturity.

For example, Wulfia requires at least one leaf: Xanthium responds well if it has few partially mature leaves. In Zea mays, at least there should be 5-6 leaves to respond for photo periodic treatment.

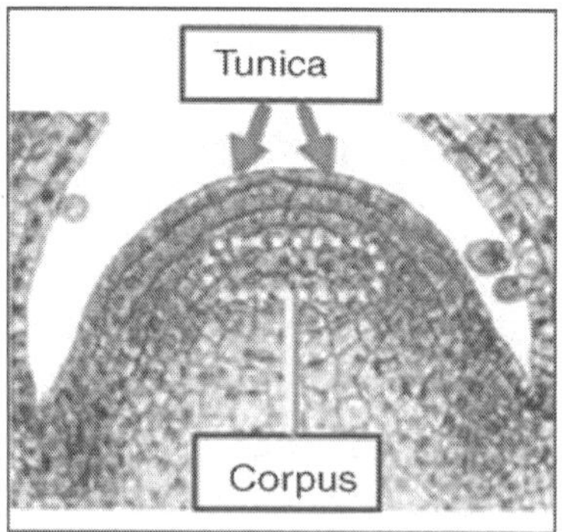

Stem apex is the site for development of flower; a dramatic change in the structural and functional features of the floral structures; they are nothing but modified leaves. In spite of it is to be noted, the floral structures are same as the vegetative structure but modified; sepals, petals, stamens and carpels are all derived from leaves.

The most intriguing question that has puzzled scientists for such a long time, how the genes hitherto remain silent, start expression and develop these structure. The molecular aspects of gene expression that changes leaves into floral organs are just opening up, yet only in bits. Looking at the start of the plant body from the zygote a fertilized egg, in course of time divides and redivides and differentiates and develops tissues and organs. Developmental process in plants, at molecular level is more or less the same pattern as in animal systems.

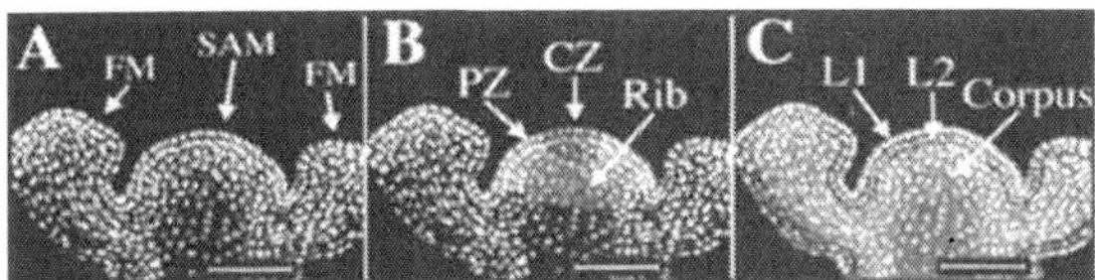

Look at the Stem Apex Meristem (SAM) has undifferentiated central dome of progenitor cells (pleuripotent in nature) covered by a layer of epidermal cells. Signals have to come from different modes and methods to convert such potent cells to go through developmental programmes. It is possible one such cell is enough for the development, just like stem cells in animal systems.

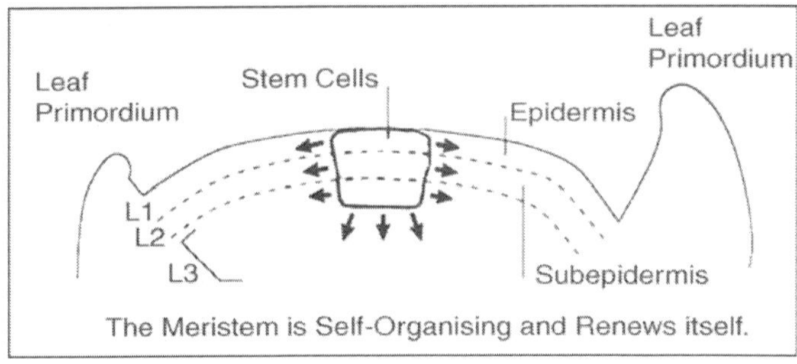

The Meristem is Self-Organising and Renews itself.

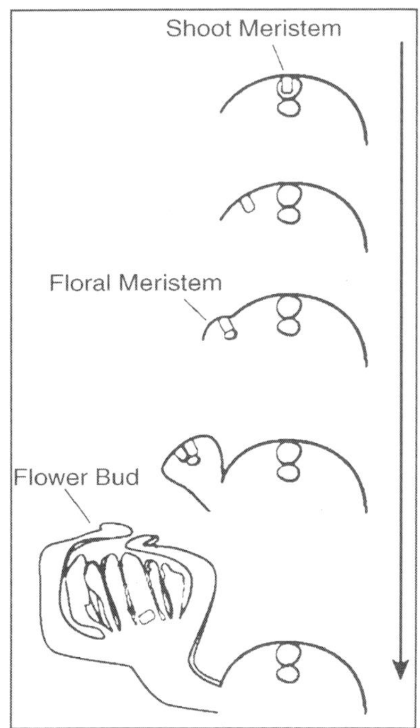

Signals are varied such as sunlight, sunlight duration, temperature, and organic chemicals such as Gibberellins and sucrose are internal signals. Cellular transduction, in response different signals, is more complex, for the signals arrive as environmental factors Light (Photoperiodic pathway) temperature (Vernalization pathway) or inbuilt factor (autonomous pathway).

Most of the environmental signals impinge on leaves and resultant downstream products have to be translocated to the SAM; this can be long distance for the site of perception and the site of response are separated in time and space (Einstein).

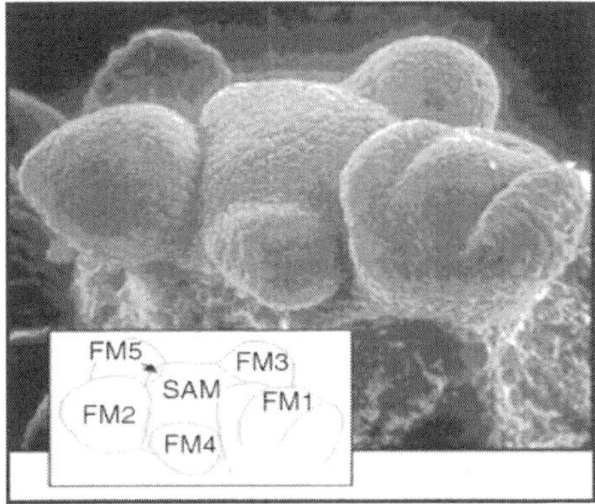

Plants without leaves do not respond to any photoperiodic treatment, which suggests that the vegetative buds are incapable of perceiving the stimulus. Similarly, plants with only old and mature leaves not only fail to respond but also they inhibit or nullify the photoperiod effect. However, partially mature leaf or leaves that are just unfolding, are highly sensitive. A remarkable feature is that even one such sensitive leaf is enough to respond to proper photoperiodic induction. It means whatever reactions or a product produced in one leaf is enough to induce flowering in the entire plant. Let us start with light mediated induction.

NON-ESSENTIAL PARTS OF THE FLOWER

The calyx and corolla whorls of floral leaves together constitute the perianth of the flower, and as they are not directly concerned in the production of seeds are termed the non-essentialparts of the flower. When one of the whorls of the perianth is absent as in the mangel, male hop, and anemone, the flower is spoken of as monochlamydeous; if both calyx and corolla are absent, as in the ash and willow, the flower is naked or achlamydeous.

THE CALYX

The calyx forms a protective covering for the rest of the flower when the latter is still young, and may either lall off when the flower opens, in which case it is caducous^ or remain attached to the receptacle for an indefinite period, when it is described as a persistent calyx. It is usually green but may assume some other colour, in which case it is spoken of as petaloid.

A calyx which consists of free separate sepals, as in the buttercup, is termed polysepalous; those in which the sepals are united, as in the primrose and pea, are said to be gamosepalous. In groundsel, thistle, and other plants belonging to the Compositae, the calyx takes the form of a ring of hair known as a pappus, which generally develop's rapidly after the corolla has faded and acts as a float for the distribution of the seed-case by means of the wind.

THE COROLLA

This part of the flower is usually of bright colour and serves mainly as an attraction for insects. When the petals forming it are free from each other, as in the buttercup and rose, the corolla is polypetalous; the term gamopetalous is applied to corollas which are composed of united petals, as in the primrose and Canterbury bell.

PLACENTATION

The arrangement of the placentas or points from which the ovules arise inside an ovary is termed placentation. When the ovules are arranged in lines on the wall of the ovary, the placentation is parietal.

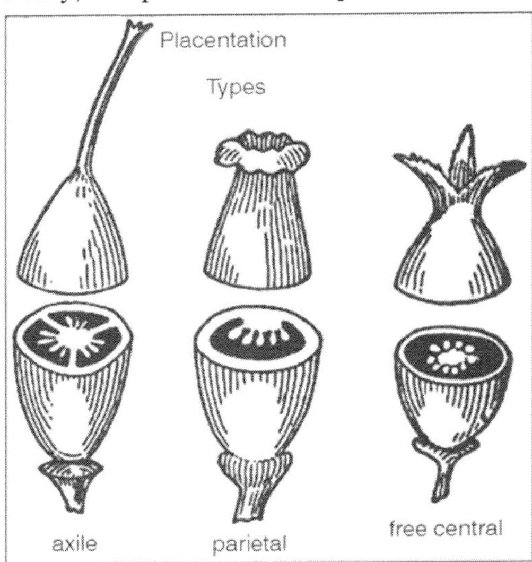

Fig. Placentation

In multilocular ovaries, such as at the ovules are generally arranged in the angles formed at the centre where the edges of the carpels are united, and the placentation is described as axile. In the primrose and chickweed families of plants the ovules are attached to a placenta which arises in the form of a short column from the base of the ovary and has no connection with the sides: this arrangement is known as free central placentation.

MONOCLINOUS AND DICLINOUS FLOWERS

When both the essential parts are present in the same flower, as in the buttercup, charlock, and the majority of common plants, the flower is described as monoclinous; sometimes the terms perfect, hermaphrodite or bisexual are applied to such flowers. In certain flowers, as those of the cucumber, melon, hop, hazel, and willow, one or other of the essential parts are missing: such are said to be diclinous imperfect or unisexual.

Diclinous flowers may be of two kinds:

- Those in which the andrcecium is alone present and described as staminate or male flowers,
- Those in which only the gynaecium is met with and spoken of as carpellary, pistillate or female flowers.

When both kinds of diclinous flowers are met with on the same individual plant, as in the case of the cucumber and hazel, the plant is said to be moncecious; in examples, such as the hop and willow where the two kinds of diclinous flowers are produced on separate individuals, the plants are spoken of as dioecious.

REPRODUCTIVE DEVELOPMENT OF PLANT

The rotation of our planet results in regular changes in environmental cues such as day length and temperature, and organisms have evolved a molecular oscillator that allows them to anticipate these changes and adapt their development accordingly. In many plants, the transition from vegetative to reproductive growth is controlled by photoperiod, which synchronizes flowering with favourable seasons of the year.

One of the most important developmental decisions in a plant's life is when to switch from vegetative to floral (reproductive) growth. If flowering is initiated at the wrong time of the year, it will affect the number of seeds produced and significantly reduce reproductive success. In addition to its role in directly initiating flowering, light also serves as an entraining signal in resetting a plant's circadian clock. It is likely that all five phytochromes regulate red/far-red entrainment signals.

The circadian system is crucial for photoperiod perception, as alterations in clock function alter the induction of flowering time. In fact, many genes that regulate clock function were originally isolated as floral-timing mutants. From work on toc1, cca1 and lhy mutant lines, a molecular model for the central oscillator was proposed. In this model, TOC1 expression in the night activates CCA1 and LHY; the resulting CCA1/LHY expression in the morning represses TOC1 expression. This repression is relieved by decay in CCA1/LHY expression, so that TOC1 levels increase again in the subsequent night, closing the 24 hour loop. New data argue against this model, however, as it appears that LHY and CCA1 are not absolutely required for circadian function. This has led to an alternative model, in which sequential periodic expression of TOC1 and its four homologs generates the oscillator. Regardless of the exact molecular details of the oscillator, it is clear that the circadian system integrates photoperiod Perception and the UV/blue-light-absorbing cryptochromes. Mutations in the red-light receptor phytochrome B result in early flowering, particularly in the absence of an inductive photoperiod. This establishes this receptor as a repressor of flowering time. Mutations in cryptochrome 2 cause delayed flowering time under inductive photoperiods.

Thus, cryptochrome 2 is a promotive receptor of flowering time. The competition between these two receptors could act in the determina-tion of repression or activation of reproductive development. Molecular genetic analyses have identified photoreceptors and light-signalling components, and components of the circadian system, which are essential for a plant to make a proper photoperiodic response. Independent studies by the Carre and Kay groups have now provided a molecular foundation for understanding how light perception is integrated with the circadian system in the generation of a photoperiodic response. Their work supports the classical 'external coincidence' model which functions through the circadian expression of the CONSTANS gene and light activation of the encoded protein.

In the external-coincidence model, a physiological response, such as flowering, is triggered when light perception coincides with the time when expression of a circadian-regulated gene exceeds a required threshold. The long days (LD), light is perceived both at dawn and dusk of the expression phase, whereas under short days (SD), threshold expression is restricted to the dark. In the internal-coincidence model the effect of light is simply to entrain two distinct circadian oscillators.

In the example shown, long days cause the two rhythms to be entrained with similar phases; this could generate two regulatory molecules that require each other's activity for physiological function. Under short days, the phases of the two entrained rhythms are further apart; this could restrict the simultaneous expression of two factors, thus inhibiting their co-action.

The link in Arabidopsis between light perception, the oscillator and flowering time appears to be the transcription factor CONSTANS (CO). Loss-of-function co mutants flower late in inductive long days, whereas ectopic overexpression of CO promotes early flowering independent of daylength.

Further, CO expression level is reduced in late-flowering gi and lhy mutants, and elevated in early-flowering elf3 and elf4 mutants. Defects in CO expression can thus explain the opposite effects on flowering time of these mutations. CO expression in Arabidopsis is modulated by light perception and the circadian oscillator, and so is dynamically regulated by daylength.

In long-day photoperiods, CO abundance is high at the beginning and end of the photoperiod, but in short days, CO is restricted to the dark phase of the day. It was thus proposed that CO expression induces flowering as a function of daylength via a light-dependent post-transcriptional process that requires a threshold level of circadian transcribed CO.

Further support for this view comes from the finding that the expression of the CO target gene FLOWERING LOCUS T (FT) is restricted to when the expression of CO is coincident with light. Although this supports an external-coincidence model for flowering time, only the requirement of light was directly examined in these experiments

When to flower is therefore a critical decision, and consequently multiple mechanisms have evolved to align flowering with optimal environmental conditions. But how does a plant recognize the presence of favourable conditions and integrate this information with its own endogenous developmental programme?

A new clue to this problem comes from the work of He *et al.* 2003. To dissect the molecular processes that initiate flowering and trigger the change from vegetative to reproductive growth, biologists have carried out intensive genetic studies of flowering time in the model plant Arabidopsi. This has led to the discovery of many genes involved in the regulation of flowering time and the development of a number of genetic models.

GENETIC PATHWAYS

Despite progress in identifying the genetic pathways involved, the mechanisms by which these flowering gene products modulate the floral transition is largely unknown. He *et al.* in 2003 elucidated one mechanism by which plants regulate flowering time.

They have identfied a protein called FLOWERING LOCUS D (FLD), which removes acetyl groups from (deacetylates) histone proteins in chromatin containing the FLOWERING LOCUS C (FLC) gene. The FLC protein encoded by this gene is a member of the MADS-domain family of transcription regulators and is a strong repressor of flowering. By deacetylating histones in FLC chromatin, FLD prevents transcription of FLC enabling the plant to flower.

To ensure that flowering occurs at the correct time in Arabidopsis, floral initiation is regulated by the integration of signals from developmental pathways and environmental cues. These include pathways that monitor plant hormones, detect light quality and duration, and respond to prolonged exposure to cold (vernalization). A number of these pathways converge on the common target FLC.

The autonomous promotion pathway regulates flowering independently of photoperiod (hence its name) by repression of FLC expression. FLD is one of six identified members of the autonomous pathway. Like other mutants in this pathway, fld mutants exhibit delayed flowering due to an increase in production of the flowering repressor FLC.

In their new work, He and colleagues show that FLD is homologous to the human protein KIAA060, a component of the human Histone Deacetylase 1,2 (HDAC 1/2) complex. HDAC complexes remodel chromatin by removing acetyl groups from lysine residues in the tails of histones. Hyperacetylated histones are associated with transcriptionally active genes, and hypoacetylated histones with transcriptionally silent chromosomal regions. To examine whether FLD may be a component of a plant HDAC complex, He and co-workers studied the acetylation state of histone H4 at the FLC locus. They immunoprecipitated specific chromatin fractions with an antibody against acetylated H4 histone tails and inspected the different fractions using polymerase chain reaction—a technique called chromatin immunoprecipitation (ChIP). They found thatin fld mutants of Arabidopsis, histone H4 tails in a specific region of the FLC locus are hyperacetylated compared with those in wild-type plants.

So, when the deacetylating activity of FLD is disrupted, FLC remains acetylated and is actively transcribed, resulting in a delay in flowering. HDACs are often recruited to their targets through interactions with cis-regulatory elements. To define these cis-elements, He and co-workers generated a series of internal deletions of the FLC gene and introduced them into plants lacking functional FLC. Removal of a 294–base pair (bp) region within intron 1 of FLC promoted hyperacetylation of FLC chromatin, FLC expression, and late flowering. Thus, deletion of this specific region of intron 1 has effects on FLC expression similar to those observed in fld mutant plants. Taken together these results show that FLD regulates FLC by deacetylating histones in FLC chromatin, providing evidence for chromatin regulation of FLC expression. The other components of the autonomous pathway also down-regulate FLC expression.

But He *et al.* demonstrate that, with the exception of FVE, they do not cause similar acetylation changes at the FLC locus. The autonomous pathway, therefore, must exploit multiple mechanisms to regulate FLC. Other components of the autonomous pathway include LD, which encodes a homeodomain protein; FPA and FCA, which encode RNA-binding proteins; and FY, which encodes a polyadenylation factor.

EXACT MECHANISM

However, the exact mechanism by which these components regulate FLC is still not known. Further studies examining how FLD regulation of FLC is integrated with information from other pathways modulating FLC will provide a useful model for understanding how regulatory networks converge on a single target. Although the homology of FLD to a component of the mammalian HDAC1/2 complex would imply that FLD is part of a similar complex in plants, this still needs to be demonstrated. Four genes with homology to HDAC1/2 have been found in the Arabidopsis genome. However, the effects of mutations in these genes do not resemble those in fld mutants.

This may reflect redundancy among the HDACs such that no one mutation alters flowering time. Consistent with this, treatment with an antisense transgene that is likely to suppress several of the HDACs does result in late flowering. The discovery of multiple FLD homologs in Arabidopsis raises the possibility that different FLD-like proteins are part of HDAC complexes with different target-site specificities.

The He *et al.* study encourages further analysis of the presumptive FLD complex, which should provide a greater understanding of the evolutionary conservation of its constituent proteins and their involvement in transcriptional repression in both plants and mammals. Such an analysis may also reveal the ways in which plant development is more plastic and adaptable than animal development. Many plant cells are totipotent, so plants need versatile mechanisms to reprogram chromatin states.

Many tools are now in hand to dissect these mechanisms and to address how they differ from those controlling reprogramming of animal genomes. The acetylation states of FLC chromatin in wild-type plants (top), fld mutants (middle), and plants lacking a specific 294-bp region of FLC intron 1 (bottom). (Top) In wild-type plants, FLD acts as part of an HDAC complex, which interacts directly or indirectly with intron 1 of FLC to deacetylate specific regions of its chromatin.

Consequently, FLC expression is reduced and the plant is able to flower. (Middle) In plants lacking functional FLD, FLD-dependent deacetylation of FLC is lost. Presumably, the HDAC complex containing mutant FLD can no longer be targeted to FLC. Therefore, the FLC locus remains acetylated, is actively transcribed, and flowering is delayed. (Bottom) Removal of a 294- bp region of FLC intron 1 also prevents FLD-dependent deacetylation of this locus possibly because the HDAC complex can no longer bind to FLC. Thus, FLC remains acetylated, and the gene is expressed, with a consequent delay in flowering.

In winter-annual types of Arabidopsis, flowering is delayed unless plants are vernalized. The delayed flowering is due to dominant alleles of FRIGIDA (FRI) and FLC. FRI elevates expression of the MADS-box transcriptional regulator FLC to levels that suppress flowering. Vernalization promotes flowering primarily by repressing FLC expression. The repressed state of FLC is maintained through mitotic cell divisions after a return to warm growing conditions. Many summer-annual accessions of Arabidopsis flower rapidly without vernalization because such accessions lack an active FRI allele or have a weak FLC allele and thus have low levels of FLC expression.

To ensure flowering in favourable conditions, many plants flower only after an extended period of cold, namely winter. In Arabidopsis, the acceleration of flowering by prolonged cold, a process called vernalization, involves downregulation of the protein FLC, which would otherwise prevent flowering. This lowered FLC expression is maintained through subsequent development by the activity of VERNALIZATION (VRN) genes. VRN1 encodes a DNA binding protein whereas VRN2 encodes a homologue of one of the Polycomb group proteins, which maintain the silencing of genes during animal development. Here we show that vernalization causes changes in histone methylation in discrete domains within the FLC locus, increasing dimethylation of lysines 9 and 27 on histone H3. Such modifications identify silenced chromatin states in Drosophila and human cells. Dimethylation of H3 K27 was lost only in vrn2 mutants, but dimethylation of H3 K9 was absent from both VRN1 and VRN2, consistent with VRN1 functioning downstream of VRN2. The epigenetic memory of winter is thus mediated by a 'histone code' that specifies a silent chromatin state conserved between animals and plants.

ROOT DEVELOPMENT

SOIL AERATION

Soil oxygen movement to plant roots is critical to maintain adequate respiration for growth. Anaerobiosis, or oxygen stress, occurs in soil when the rate of supply falls below the biological demand. Impeded soil aeration can arise from poor drainage and waterlogging or from compacted soil. One approach to evaluate soil aeration is to measure the composition of soil air. This technique indicates that gas exchange between the atmosphere and soil air is restricted when oxygen content of the soil air falls significantly below that of the atmosphere. An important challenge with this approach is how to extract a representative sample of soil air that avoids mixing of gases from outside the point of sampling or contamination from the atmosphere.

Another approach for assessing soil aeration is to determine the air porosity, or fractional air space, at a standard soil water potential. Work to identify threshold air porosity values below which plant growth is limited has generated values ranging from 5 to 20 per cent.

Madison has indicated that 10 per cent air porosity is the threshold level for intensively utilized turf. The United States Golf Association Green Section (USGA) has suggested 15 per cent air porosity at a 3 kPa water tension as the lower test limit for materials used in the construction of root zones for golf course putting greens. The extent to which air porosity in the root zone changes with time under a perennial turf system is a subject of current studies.

The rate of air exchange, rather than the volume of soil air, is the more important factor in soil aeration and therefore limits the usefulness of air porosity as an index of soil aeration. Gaseous movement in the soil occurs by two main processes: convection and diffusion. Convection, where the moving force is a gradient of total gas pressure, is thought to be a minor mechanism of soil aeration except at shallow depths and in soils with large pores. Although convection is not considered an important factor for more deeply rooted plant species, it could be more influential for shallow-rooted turfs. A large fraction of the root system of moderately to highly maintained turfs is found at surface depths of 5 cm or less. Furthermore, the crown, the major meristematic organ, of turfgrass plants is located near the soil surface. Thus, turfs with a high plant population could have sufficiently high oxygen demands at the soil surface that convection would play a useful role in maintaining adequate aeration for plant crowns and surface rooting.

Convection, therefore, could be partially responsible for the observations of improved turfgrass quality after changes in air masses (pressure) associated with passing weather systems. Below the surface, however, diffusion is considered the more important exchange mechanism in soil, and the moving force is a gradient of partial pressure.

Gas exchange between the atmosphere and soil is maintained predominantly by diffusion through air-filled pores. In contrast, live tissue is typically hydrated, and the supply of oxygen to roots occurs by diffusion through water films. This is an important distinction, since the diffusivity of oxygen in water is less than its value in air by a factor of 10"4.

The solubility of oxygen in water may be a compounding issue, especially at higher temperatures, where solubility decreases. Methods that measure gaseous diffusion through air-filled pores do not provide information on the impedance to oxygen presented by the water film surrounding a root. It is possible that an inadequate supply of oxygen to roots can occur in a well-aerated soil (*i.e.*, high air-filled porosity and O_2 concentration) if the roots are surrounded by thick water films.

Thus, under such conditions, measurement of soil oxygen movement to plant roots would provide a more complete assessment of soil aeration. Lemon and Erickson introduced the method commonly called ODR (oxygen diffusion rate) to measure oxygen diffusion to a root-like electrode inserted into the soil and enveloped in water films of the soil. The ODR technique has been used by a number of researchers to demonstrate the reduction in oxygen supply in compacted soil. Limiting soil aeration conditions are often transient and strongly affected by wetting and drying cycles in compacted soil.

Therefore, the timing of gas composition and ODR measurements is critical for accurate detection of anaerobiosis. Generally, 200 ng cm^{-2} min^{-1} has been considered the minimum ODR value, below which plant roots will not grow. Some researchers have reported values ranging from 50 to 200 ng cm^{-2} min^{-1} as limiting ODR values for turfgrass root growth. The interpretation of these results has been questioned because, in addition to the diffusion of oxygen, other soil factors and components could have affected the measured ODR.

Work by Blackwell has helped to reduce errors associated with reactions other than the reduction of oxygen in the ODR technique.

Moreover, early studies of limiting ODR were performed in growth chambers at about 20æ per cent C, whereas soil temperatures in the field can vary considerably during the growing season. Blackwell and Wells concluded that the effect of oxygen flux on root growth is related to temperature differences and differences in root respiration. Thus, it is plausible that the oxygen flux limiting root growth under high soil temperatures may be greater than values that have been observed at 20æ per cent C.

When oxygen becomes limiting, some species of bacteria shift metabolic pathways so as to utilize other compounds as terminal electron acceptors. Thus, plants may have not only to survive periods without oxygen but also to withstand toxic substances such as hydrogen sulfide, which are produced during anaerobic microbial respiration.

Development of aerenchyma tissue (intercellular spaces) that allows diffusion of oxygen from aerial plant tissue down to roots may be an adaptive response that helps certain species of plants to survive periods of soil anaerobiosis. Agnew and Carrow reported increased root porosity in Kentucky bluegrass (Poa pratensis L.) grown under compacted soil conditions. They observed the greatest root porosity under conditions of both compaction and water stress.

SOIL STRENGTH

Soil strength can be described as the capacity of soil to resist a force without rupture, fragmentation, or flow. Various methods exist to quantitate soil strength in the laboratory, but the method most used in the field is the penetrometer. It measures the effort necessary to push a thin, cone-tipped probe into the soil.

Taylor and colleagues conducted pioneering research demonstrating that the strength of a soil, as measured with a static penetrometer, will increase as soil bulk density increases and water content decreases; consequently, mechanical impedance to root growth is greater. The limiting strength of soil to cotton (Gossypium hirsutum L.) root growth is in the range 2.5 to 3 MPa. For many crops, a 2-MPa penetrometer resistance is commonly encountered under field conditions and can reduce root length and elongation by at least 50 per cent.

ROOT MORPHOLOGY

In experiments designed to study the constricted growth of roots, the ability of ryegrass (Lolium perenne L.) roots to enter rigid pores (glass capillaries and sheets of steel) depended on the size of the root cap and the thickness of the stele, which needed to be about one-third of the nominal root thickness. Roots could elongate down long capillaries while constricted, albeit at a reduced rate. Roots can enter pores of diameter smaller than the root tip itself only if the rigidity of the pore structure is weak enough to allow soil displacement. Thus, major root axes typically thicken when soil pores are too small for root penetration. Ethylene production has been linked to this thickening response of roots subjected to mechanical impedance. Abscisic acid, which increases temporarily under conditions of mechanical impedance, has also been shown to induce root thickening as well as increased root-hair number and curling of roots. The thickening of the root compresses soil laterally and minimizes friction with soil as the root extends axially. Additionally, root-hair development aids in anchoring the root as it penetrates compacted soil. Thus these responses can be viewed as adaptive, as they mitigate the effects of axial resistance on growth. Additional contributing factors to thicker roots' relative endurance of compact soil might be greater resistance of thicker roots to bending or higher axial pressures exerted by thick roots.

Variation in plant genetics exists for tolerance to high mechanical impedance. Interspecific comparisons have shown differences in root dimensions and the ability to penetrate soils. Greater root thickness and the tendency for roots to expand radially in response to mechanical impedance are correlated with the capacity to elongate in hard soil.

Intraspecific studies have also demonstrated genetic variation in characteristics associated with tolerance to mechanical impedance. Two tall fescue (Festuca arundinacea Schreb.) lines with large-diameter roots were able to penetrate a hardpan at the 0.4 to 0.6 m soil depth, extracted more soil water from the 0.6 to 1.2 m depth, and yielded 40 per cent more in dry matter compared to fescues with small-diameter roots.

Development of efficient techniques to screen plant germplasm for greater tolerance to high mechanical impedance could prove to be highly useful. Iijima *et al.* described the morphological development of rice (Oryza sativa L.) and maize (Zea mays L.) root systems as affected by soil compaction. Elongation of the main root axes of rice and maize was restricted in compacted soil.

Seminal roots of maize were much less restricted than those of rice. The root system of rice was characterized by a larger fraction of long, thick laterals with potential to produce higher-order laterals on their axes. Growth of higher (second- and third-) order lateral roots in compacted soil compensated for the restricted growth of the main root axes in both species. The work of Carrow and colleagues indicates that the effects of compacted soil on root growth varies with plant species and management. Total root mass of Kentucky bluegrass decreased in response to compaction treatment, whereas tall fescue and perennial ryegrass root systems were more tolerant to compaction stress.

Subsequent work indicated that the most detrimental effect of compaction on root growth of tall fescue and perennial ryegrass occurred under higher nitrogen fertilization. Although total root mass of a turf may not change under compacted conditions, a redistribution of the root system can occur where root mass increases in the surface 5 cm of the soil at the expense of rooting below the 10-cm depth. Root mass may not provide the most detailed index of a root system's response to compacted soil, since root mass is affected less by bulk density than root length is, primarily because of larger-diameter roots in high-density soil. Thus, the measurement of the number and length of roots growing in soil may be a more sensitive indicator of compaction stress on a root system than root mass. Root length and number measurements, however, are expensive, particularly for experiments evaluating a large number of treatments and for plant systems such as turf, which have extremely high root lengths.

Physical alteration of water and nutrient uptake by a root system can result as the orientation and morphology of roots change. Higher soil bulk density brings more soil, and associated water and nutrients, into contact with the surface area of the root system and thus has the potential to increase availability.

This effect is particularly pronounced for nutrients such as phosphorus, which have relatively low mobility in the soil. Greater soil contact with the root system, and thus availability of nutrients and water, could at least partly explain the greater clipping yields of turf growing on heavily compacted plots compared to lightly compacted plots during the spring. The fact that this response was reversed during the summer suggests that the beneficial response of greater soil and root surface contact may be beneficial until environmental conditions become limiting for other essential growth factors under compacted soil conditions. As environmental conditions become more limiting, the reduced total root elongation rate and length in compacted soil ultimately decreases availability of water and nutrients by restricted access.

SHOOT RESPONSES

The direct and indirect effects of compaction on plant roots can elicit responses in the plant shoot as well. A number of studies on grasses demonstrate the potential of compacted soil to limit shoot density, verdure, clipping yield, soil cover, and lateral stem growth.

VEGETATIVE ANATOMY OF THE FLOWERING PLANTS

The vegetative anatomy of the flowering plants, allowing for some variation, is remarkably consistent form one plant to another. The cell and tissue types are almost identical to the ferns. The flowering plant embryos differentiate early into three types of dividing or meristematic regions. These are outside protoderm, bulk ground meristem and the area that will develop into vascular tissue, the procambium. These embryonic tissues or primary meristems will in turn differentiate into three basic tissue systems.

These tissue systems are, outer or dermal system, the vascular tissue system and the remainder of the plant or the ground (fundamental) tissue system. Each of the plant organs has these tissue systems arranged in a particular fashion and comprised of a particular group of cells. The cells and the arrangement depend on the different functions of the organs to the plant as a whole.

THE LEAF

The leaf structure is a compromise between three conflicting, environmentally induced, evolutionary pressures. These are: the need to expose the maximum photosynthetic surface to sunlight; the necessity of conserving water; and the need to have an adequate gas exchange between the inside and outside leaf environment to permit photosynthesis. If we deal only with regular photosynthetic leaves. They appear to be separated into three distinct regions, the epidermis, the mesophyll and the veins or vascular bundles.The epidermis is similar to that of the ferns. The tissue that makes it up is epidermal and the cell type is also termed epidermal.

The same modifications of cuticle and guard cells exist for the same reasons. The mesophyll, made of fundamental tissue and chlorenchyma cells, is usually more frequently organized into distinct palisade and spongy layers than is found in the ferns. The vascular bundles are similar to those of the ferns, but three new cell types appear in the vascular tissue of the flowering plants. In the ferns, the xylem was made of conducting elements called tracheids, while in the flowering plants there is also another type of conducting unit termed the vessel. Vessels begin their development in a similar fashion to the tracheids, but at maturity, they lose their end walls. This forms a continuous tube so the water and minerals do not have to travel in a zig-zag fashion but rather can travel straight through the leaf. These tubes of vessels are continuous with the vessels in the stems and roots and provide a more rapid and efficient route of transport. The phloem of the flowering plants is organized in a similar fashion to the vessels of the xylem. Instead of sieve cells, the phloem contains sieve tube elements or cells. The difference between these and the sieve cells is again because of the perforation of the end walls and not a solid end plate with sieve units on the lateral walls.

The sieve tubes are anucleate at maturity but they still contain a living functional cytoplasm. The control of the activity of this cytoplasm is brought about by a group of connected adjacent cells termed companion cells. These tiny, elongate cells direct the sieve tube activity through plasmodesmatal connections. The flowering plants do not have the sieve cells. Vascular bundles of the leaves are usually surrounded by a special, single cell layer termed a bundle sheath. The parenchyma cells of this sheath are responsible for transferring large amounts of glucose that has been produced in the leaf mesophyll, to the phloem.

THE STEM

In many respects, the stems of the flowering plants are similar to the stems of ferns. The variations come partially from the need to adapt to above ground conditions and partially from the need to adapt to the wider range of habitats occupied by the seed plants. There are two main patterns of organization in the stem anatomy. The dicotyledons present one anatomical pattern while the other is found in the monocotyledons. The monocotyledons stems have only three regions, an epidermis, cortex and vascular bundles scattered throughout the cortex. The dicots have four regions. The epidermis, cortex, a ring of vascular bundles and a central pith.

The epidermis is different from that of the ferns since it now must protect the stem from the hazards of an above ground environment. Also, the stems of these plant are frequently photosynthetic so guard cells are also required.

The cortex varies according to the plants and the environment. Plants in drier regions often have many sclerenchyma cells in their cortex, while those of more moderate areas have parenchyme (chlorenchyma) and collenchyma. Collenchyma cells are living, strengthening cells which have increased cellulose fibres in the walls. These cells are usually found at the outer surface of the stems, frequently in the corners.

The pattern of organization of the vascular bundles of the dicots is a ring. The vascular tissue is again subdivided into xylem and phloem and these two subtissues have the same cell types as in the leaves of the flowering plants. The xylem of the vascular bundles always is located on the inside of the phloem, but the phloem never rings the xylem as in the fern rhizome vascular bundles.

The scattered bundles of the vascular tissue of the monocots has proportionately more vessels than that of the dicots.The pith, found only in the dicots, comprises the central region of the stem and is made up of parenchyma cells. These cells, rich in leucoplasts, are adapted to store the products of photosynthesis and water.

THE ROOTS

Roots have the least cell and tissue differentiation of all the organs. They function in absorption, storage, transport and anchoring. The absorption properties are found only in the extreme tips of the roots and root branches. After approximately 10 - 20mm from the tip, the root tissues mature and water and minerals can no longer be absorbed. The root epidermis develops a cuticle to prevent loss of the water already absorbed and to reduce the affect of harmful soil organisms.

At the extreme tip of the roots is a layer or cap of constantly replaced dead cells. This root cap protects the root as it grows through the soils in search of sufficient water and minerals. Behind the root cap is a zone where water is absorbed. To facilitate this process, many of the epidermal cells in this region develop projections out from their outer walls termed root hairs. The root hairs are lost after the area of absorbtion in replaced by more mature tissues.

The outer region of the roots is again the epidermis.The cortex occupies the bulk of the root. It is often quite well developed with large numbers of parenchyma cells. These cells contain many leucoplasts and are the site of starch storage. Biennial plants store the nutrients for over winter in this region. The inner layer of the cortex is the Casparian strip. This layer functions in a similar fashion to the comparable layer of the fern rhizome.

The vascular bundle or stele of the root is located in a central position. The outer cell layer, termed the pericycle is formed from relatively undifferentiated parenchyma cells. These cells can easily revert back to meristematic or dividing cells.

If a root branch is required, it originates from this layer. This allows the newly forming vascular tissue of the branch to link up with the vascular tissue of the primary root and eventually form part of the continuum of the vascular system that runs throughout the entire plant. The xylem is again central to the phloem, as it was in the stem. The cell and tissue types are the same.

ANATOMY OF STEMS

Since, the mechanism of the movement of water can scarcely be understood intelligently without some knowledge of the anatomy of the tissues through which it moves, a brief review of stem structure is desirable before proceeding further with a discussion of this process. Stems vary greatly in their structure, every species possessing some anatomical features which are peculiar to itself. Nevertheless certain general patterns of tissue arrangement have been found to prevail, and the stem structure of most species will approximate one or another of these general arrangements.

The corn stem represents an herbaceous monocot type of structure while that of the tulip poplar is typical of woody dicot stems. The structure of woody stems cannot properly be appreciated merely from a consideration of one year old stems. The primary tissues of such stems develop from tissues formed at the apical growing tip during growth of the stem in length. Nearly all perennial stems also grow in diameter as a result of the development of secondary tissues from the cambium.

By successive stages of division, enlargement, and differentiation of cambium cells additional layers of xylem (secondary xylem) are laid down on the inner face of the cambium, and new layers of phloem tissue (secondary plzloem) on its outer face Hence, after a few years the great bulk of any woody stem is composed of secondary tissues. Secondary xylem and secondary phloem also develop from the cambium in most herbaceous dicot stems but in such species the cambium in any one of phloem and xylem elements from the cambium.

- A cambium initial,
- Division of cambium cell,
- Outer cell resulting from division becomes a phloem element,
- Another division of the cambium cell,
- Outer cell resulting from second division becomes a xylem element.

Actually several xylem or several phloem elements are often formed successively. Upon maturation most of the xylem and phloem elements acquire sizes and shapes which are very different from those of the cambium initials from which they originate. Stem is never active for more than one growing season.

The spring formed xylem tissue, as viewed in cross section is usually distinctly different in aspect from that formed later in the season. In many angiosperms the springwood contains more and larger vessels and the cell walls are generally thinner than in the subsequently produced summer wood. In the conifers the spring formed tracheids are thinner-walled and of larger cross-sectional diameter than those formed later in the growing season.

The transition from spring to summer wood is often a very gradual one. On the other hand the more open xylem tissues formed each spring abut directly upon the denser tissues which were produced during the preceding summer, thus giving rise to an abrupt line of demarcation between the zones of xylem formed in any two successive seasons.

The result of this growth behaviour is that a cross section of the trunk or a branch of any tree appears as a system of concentric layers, the so-called annual rings, each representing an annual increment of growth. In rare cases no annual rings nor more than one annual ring may be produced in a season, but usually each ring represents the xylem produced by the activity of the cambium during one season.In many woody species as the xylem tissues increase in age important changes occur in the colour, composition, and structure of the various elements, resulting in the conversion of sapwood into heartwood.

As sapwood ripens into heartwood the walls of any remaining living cells of the xylem become increasingly lignified death of these cells soon following. The water content of the tissues is generally reduced and such compounds as oils, resins gums, and tannins accumulate in the cells or cell walls. The darker colouration of the heartwood of most species as compared to the sapwood is due to such accumulations. In mature trees the heartwood becomes merely a central supporting column surrounded by a cylinder of sapwood which varies in thickness from a few to many annual layers, depending upon the species and the environmental conditions under which the tree is growing.

In some species (apple, elm) the heartwood remains virtually saturated Tangential section of a small portion of the wood of whitepine (Pinus strobus). The vertically oriented elongated elements are tracheids. Bordered pits show in sectional view in the tracheid and marginal ray cells. Simple pits show in sectional view in other ray cells. Half-bordered pits show between marginal ray cells and other ray cells.

The water in the heartwood of such species as apple and elm appears to be largely static and is not directly involved in translocation. Coincident with the development of secondary xylem, secondary phloem tissues also develop from the cambium. Cork cambiums are also initiated in the bark which produce cork layers. Profound modifications therefore occur in the outer tissues as well as in the xylem of woody stems as they grow older.

A somewhat more detailed concept of the structure of the cells and elements of the stems of angiosperms through which the movement of water occurs is presented which represent longitudinal tangential and radial sections from the xylem of a tulip tree which may be taken as representative for this group of plants.

The xylem tissues of the wood of angiosperms are composed of vessels tracheids, fibres of several types, wood parenchyma, and xylem ray cells. There is a tremendous variability in the proportional distribution and arrangement of these tissues according to species.

The most characteristic elements of the xylem tissue of angiosperms are the vessels. These are, in general, more or less tubular structures which may extend through many feet of the xylem. In some species cross walls, usually perforated, are of frequent occurrence in vessels, in others such cross walls are infrequent or lacking. In diameter they may range in trees from about 20 to about 400. In vines they may be as much as 700 in diameter. The vessels branch extensively in certain regions of the plant, especially at nodes within the leaf lamina, and at the points in the root systems where rool branching occurs.

The vessels of the proto xylem (the first cells of the xylem to mature during the ontogeny of a growing stem or root tip) have cellulose walls which are reinforced by distinctive lignified thickenings which appear as rings, spirals,or other characteristic patterns.

The vessels that develop later in the ontogeny of any growing tip have lignified, usually pitted walls which lack any of the thickenings which distinguish the walls of the vessels of the protoxylem.

Pits normally occur in all parts of vessel walls which are contiguous with other vessels or cells. These pits consist essentially of thin areas in the walls. The walls of the vast majority of plant cells are pitted. The architecturally distinct type known as the bordered pit.

GRASSES

At first sight a grass plant can seem to be rather basic and primitive. However, the grasses are actually quite advanced. There are many different types, each specially adapted to survive in certain conditions. And each has its own particular community of plants and animals associated with it.

Grass plants can be annual or perennial, and are well adapted to living in difficult conditions. Some can survive droughts, others thrive in wetlands and some withstand extremes of temperature. Annual varieties are very good at colonising land that suddenly becomes available. They grow very rapidly, and within one year produce masses of seed, and then the plants die. Perennial grasses continue growing from year to year. As well as producing seed, they also spread by means of underground stems. In this way the plant can spread across an area, forming a dense mat. This is important in binding soil together and preventing erosion.

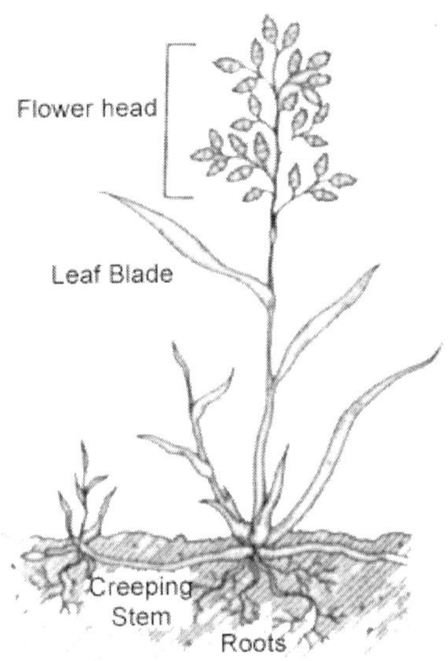

Grasses are flowering plants, and belong to the group of plants known as the monocotyledons. These plants only produce one seed-leaf on germination. The plants which we think of as wild flowers belong to the group called dicotyledons, producing two leaves on germination. Grass roots are tough, fibrous and form on clumps or underground mats. Apart from anchoring the plant and absorbing water and minerals, the roots hold the soil together, protecting it from drying out and blowing away, or becoming muddy and washed away. The stem is cylindrical, hollow and has swollen, solid joints at intervals along it. These are called *nodes*.

Each leaf arises from a node on the stem, and then tightly encircles the stem. The way that the stem of a grass grows is part of the reason why grasses are so successful. Some plants form mats, spreading either over the surface of the soil by leafy, creeping stems known as *stolons*, or through the soil by tough scaly stems known as *rhizomes*. One of the main differences between the similar grasses, sedges and rushes lies in the nature of their stems. Rushes usually have pithy stems without solid nodes, and sedges usually have 3-sided, solid stems.

The leaves of different grasses vary in shape and texture, but all grasses have elongated or elliptical leaves which have parallel veins and no stalks. The leaf is really in three parts: the *sheath*, which encloses the stem; the *blade*, or flat, green part; and the *ligule*, where the blade bends away from the leaf. In most plants the main growing point is at the tip of the stem. If this is damaged or cut off, the plant can only grow from buds further down the stem, and this takes some time. In grasses, however, the growing point is near the base of the leaf. If the top of a grass's leaf is cut off, it still keeps growing from the bottom. So when grass is eaten by animals or mown, the grass plants are not killed, but grow even more.

Both monocot and dicot plants are able to reproduce asexually, producing runners, suckers, bulbs and corms which give rise to individuals identical to the parent. This is useful as it allows a plant to spread rapidly, but it is also essential that they have the ability to reproduce sexually. This is how genes are exchanged between plants of the same species, creating genetic variation within a species, the key process behind evolution. Occasionally a variation will be produced that is more successful, and the advantageous gene will be spread through the population. It is this process that enables a species to adapt to changes in the environment. Genetic variation can also help to check the spread of disease, as some members of the species may be disease resistant. One of the problems with genetically uniform crops is that they are very susceptible to disease.

Bibliography

Alfred Barton Rendle: *Flowering Plants and Their Classification (2 Vols-Set)*, Biotech Books, New Delhi, 2005.

Alka Awasthi: *Plant Geography and Flora of Rajasthan with an Introduction to the Plant Geography of India*, Deep & Deep Publication, Delhi, 1995.

B.L. Bhellum and Rani Magotra: *A Catalogue of Flowering Plants of Doda Kishtwar and Ramban Districts : Kashmir Himalayas*, Bishen Singh Mahendra Pal Singh Publication, Dehra Dun, 2012.

B.P. Uniyal, J.R. Sharma, U. Choudhery and D.K. Singh: *Flowering Plants of Uttarakhand : A Checklist*, Bishen Singh Mahendra Pal Singh Publication, Dehra Dun, 2007.

C.K. Naidu: *Plant Growth and Development*, Pointer Publication, New Delhi, 2004.

C.S. Patil, Ajit K. Gangawane and Z.G. Badade: *Plant Mitochondria and Gene Expression*, APH Publication, New Delhi, 2012.

Dhan Singh: *The World Weeds : Quarterly Journal of Weeds and Parasitic Flowering Plants, Volume 6, Number 1-4*, International Book Distributors, Delhi, 1999.

Dipak Kumar Mishra & N P Singh: *Endemic and Threatened Flowering Plants of Maharashtra*, Botanical Survey of India, 2001.

Don Grierson: *Plant Genetic Engineering*, Springer Publication, London, 2007.

Edwin M. McPherson: *Plant Location Selection Techniques*, Jaico Publishing, Jaipur, 2003.

Ellison Banks Findly: *Plant Lives : Borderline Beings in Indian Traditions*, Motilal Banarsidass Publication, Delhi, 2008.

George Benthem: *Handbook of the British Flora : A Description of the Flowering Plants and Ferns*, Biotech Books, Chennai, 1997.

J C Willis: *A Manual and Dictionary of Flowering Plants and Ferns (2 Vols-Set)*, Cosmo Publication, Delhi, 2006.

J.G. Baker: *Flora of Mauritius and The Seychelles : A Description of the Flowering Plants and Ferns of Those Islands*, AES Publication, Chennai, 1999.

Keshab Raj Rajbhandari and Suraj Ketan Dhungana: *Endemic Flowering Plants of Nepal : Part 2*, Department of Plant Resources, 2010.

M.S. Bista, M.K. Adhikari and K.R. Rajbhandari: *Flowering Plants of Nepal (Phanerogams)*, Department of Plant Resources, National Herbarium and Plant Laboratories, 2001.

Mamatha Rao: *Microbes and Non-Flowering Plants : Impact and Applications*, Ane Books, Delhi, 2009.

N Sasidharan: *Biodiversity Documentation for Kerala, Part VI : Flowering Plants*, Kerala Forest Research Institute, 2001.

N Sasidharan: *Flowering Plants of Kerala : A Checklist (CD)*, Kerala Forest Research Institute, 2006.

N. Sasidharan: *Flowering Plants of Thrissur Forests (Western Ghats, Kerala, India)*, Scientific Publication, Jaipur, 1996.

P C Trivedi: *Plant Morphology and Biotechnology : Festschrift in Honour of Prof C M Govil*, Aavishkar Publication, Delhi, 2007.

Philip Mathew and M. Sivadasan: *Diversity and Taxonomy of Tropical Flowering Plants: Prof. K.S. Manilal Commemoration Volume*, Mentor Books, Ireland, 1998.

R. Ranjan, S.S. Purohit and V. Prasad: *Plant Hormones : Action and Application*, Agrobios Publication, Jaipur, 2008.

R.P. Kaul: *Plant Metabolism*, Swastik Publication, Delhi, 2009.

Raghavan: *Developmental Biology of Flowering Plants*, Springer Publication, London, 2005.

Ronald Good: *Flowering Plants and Their Evolution*, Agrobios Publication, Jaipur, 2007.

Suchitra Ghosh: *Records of the Zoological Survey of India: Urban* Biodiversity of Kolkata : Flowering Plants Butterflies Birds and

T.S. Nayar, A. Rasiya Beegam, N. Mohanan and G. Rajkumar: *Flowering Plants of Kerala : A Handbook*, Tropical Botanic Garden and Research Institute, 2006.

V. Raghavan: *Developmental Biology of Flowering Plants*, Springer Publication, London, 2008.

V.T. Antony, A.N. Henry and Antony Kadavil: *Biodiversity of Flowering Plants*, Sonali Publications, Delhi, 2011.

Verne Grant: *Genetics of Flowering Plants*, Bishen Singh Mahendra Pal Singh, Dehra Dun, 2004.

Y.P.S. Pundir: *The World Weeds : Quarterly Journal of Weeds and Parasitic Flowering Plants*, IK International Publication, Delhi, 2003.

Index